D1647660

S. F. BARKER graduated from Swarthmore College and then studied philosophy at Harvard University, where he earned his doctorate in 1954. During the academic year 1955–56 he held the Santayana Fellowship in philosophy at Harvard and completed this book. He has taught at the University of Southern California, the University of Virginia, and the Ohio State University, and also has served as a visiting faculty member at the University of California in Berkeley, at Harvard, and at Swarthmore. Since 1961 he has been professor of philosophy at The Johns Hopkins University.

INDUCTION AND HYPOTHESIS

A Study of the Logic of Confirmation

CONTEMPORARY PHILOSOPHY

General Editor
Max Black, Cornell University

Editorial Committee
Charles A. Baylis, Duke University
William Frankena, University of Michigan
Morton G. White, Harvard University

Induction and Hypothesis

A STUDY OF THE LOGIC

OF CONFIRMATION

By S. F. Barker

THE JOHNS HOPKINS UNIVERSITY

Cornell University Press

ITHACA AND LONDON

This work has been brought to publication with the assistance of a grant from the Ford Foundation.

The law of reason which requires us to seek for this unity is a necessary law, since without it we should have no reason at all, and without reason no coherent employment of the understanding, and in the absence of this no sufficient criterion of empirical truth. In order, therefore, to secure an empirical criterion we have no option save to presuppose the systematic unity of nature as objectively valid and necessary.—KANT, *The Critique of Pure Reason*, A 651

Preface

PEOPLE who have never stopped to think about the matter often are inclined to suppose that it is perfectly clear under what circumstances evidence confirms hypotheses. Indeed, even philosophers have begun to perplex themselves about the details of this matter only in fairly recent times. This book aims to suggest that the notion of confirmation really does involve some rather puzzling and philosophically important problems, and it aims to suggest—though very tentatively—some partial answers to these problems about confirmation.

The preparation of this book was made easier by the generosity of Harvard University in granting me the Santayana Fellowship in Philosophy during 1955–1956, and it is a pleasure for me to record my gratitude. I owe my best thanks to Professor Donald Williams and to Professor Carl G. Hempel, both of whom read an earlier version of this manuscript, made helpful suggestions, and encouraged me with regard to it. Professor Nelson Goodman has been kind enough to explain

Preface

several important points to me. And I must thank Professor Israel Scheffler for reading the whole manuscript and offering valuable criticisms. However, none of these gentlemen is responsible for my errors—as they themselves would perhaps be the first to insist. Finally, I must thank the General Editor of the Contemporary Philosophy series for his consideration and Mr. Marshall Cohen for helpful assistance.

Charlottesville   S. F. BARKER
March 1957

Analytical Table of Contents

Analytical Table of Contents

Analytical Table of Contents

Three: Eliminative Induction

Four: Enumerative Induction

Analytical Table of Contents

Five: Induction and Simplicity

Six: Reductionism

Analytical Table of Contents

Analytical Table of Contents

Eight: The Method of Hypothesis

Analytical Table of Contents

Nine: Simplicity and Confirmation

Analytical Table of Contents

Ten: Concluding Remarks

INDUCTION AND HYPOTHESIS

A *Study of the Logic of Confirmation*

One

Introductory

I

EACH one of us holds a great many beliefs, on a wide variety of subjects. Of some of our beliefs we are clearly aware, of others we are perhaps scarcely conscious; some beliefs are tentative, about others we feel very confident; some have to do with scientific facts, others just with matters of common knowledge. Most of us believe that the sea is salty, that the moon is not green cheese, that the sun will rise tomorrow, that Moses lived before Mohammed, that gold dissolves in aqua regia, that hydrogen molecules contain two atoms, that five and seven make twelve, that either it's snowing or it isn't, that Vermeer is better than Ter Borch, that one ought not to torture infants with red-hot irons.

Among our beliefs some are reasonable and some are foolish. If we are intellectually critical, if we wish to be able legitimately to claim that we *know* the things we most confidently believe, then we shall continually be examining the array of statements to which we subscribe, seeking to winnow out those that it is foolish to believe; and we shall seek also to add whatever new statements it is reasonable to believe. Yet sometimes we become perplexed about how to decide what we ought to believe. What really are the marks of a reasonable belief? That is, what basic criteria ought one to use if one is to distinguish correctly between foolish, unreasonable beliefs and sensible, reasonable ones? Traditionally, that branch of philosophy called epistemology, or the theory of knowledge, has undertaken to consider what answer, if any, this question has.

We may remark at once that many (though perhaps not all) of the statements we believe are empirical—in the sense that if we do have any reason for believing them this must involve appeal to what has been seen or heard or felt or smelled or tasted: experience is involved. Thus, suppose someone asks you why you believe that the sea is salty. If you really do possess any good reasons for holding this belief, they will have to be reasons based somehow on your experience: you might reply that you have tasted the sea, moreover that you have seen books which assert this fact, and that you have heard people say it is so. If you had not had any experiences such as these, then you would not have any reason whatever for believing the statement in question. Indeed, the bulk of our everyday knowledge as well as most of science clearly involves empirical statements, statements that are dependent upon experience in this way.

Moreover, of all the empirical statements that we believe, few really interesting ones are conclusively verified by experience. Most of the empirical statements that are important and interesting to us (generalizations and predictions, for instance) have the logical character of conjectures, or hypotheses: they may be supported or confirmed or made probable by the evidence which experience provides, but they are not rendered absolutely certain—the evidence does not logically imply them. The evidence which we possess may be very strong indeed in favor of the hypothesis that gold always dissolves in aqua regia or the hypothesis that the sun will rise tomorrow; nevertheless, it remains logically possible that these hypotheses might actually be false, in spite of the evidence we possess; further experience might yield new evidence which would oblige us to give up our belief in these statements.

Our empirical knowledge, then, may be regarded as a fabric of hypotheses, each confirmed to a greater or lesser degree by the evidence which experience provides. Now, both in science and in everyday thinking we are concerned to evaluate the soundness of arguments which purport to show that particular hypotheses are supported by particular evidence. These we may call nondemonstrative arguments, in order to contrast them with the demonstrative arguments of which deductive logic treats. A valid demonstrative argument is conclusive: if all men are mortal and Socrates is a man, then it follows conclusively that Socrates must be mortal. Here it is impossible that the premises should be true and the conclusion false. In a nondemonstrative argument the situation differs, however: if many crows have been observed and all observed to be black, then, in the absence of contrary evidence, it follows with probability that all crows are black. Here it is possible that the premises might be true yet the conclusion false—though that is improbable, and therein consists the force of the argument.

The term 'nondemonstrative' here is intended to embrace all those modes of argument which are such that the conclusion, though supported or confirmed or made probable by the premises, is not implied or entailed by them. Someone may suggest, however, that the term 'inductive' is preferable; indeed the latter term is preferred by most writers who have dealt with these subjects. Certainly when we are speaking loosely there is no harm in using the word 'inductive' in this broad sense. But actually the word 'inductive' has a more specific meaning which should be remembered when we want to speak carefully: it refers to arguments which employ premises containing information about some members of a class in order to support as conclusion a generalization about the whole class (or a prediction about an unexamined member of the class). If we limit ourselves to arguments whose conclusions are empirical statements, then it is clear that all inductive arguments are nondemonstrative; but it is by no means obvious that all legitimate nondemonstrative arguments need be inductive—that is an open question. For the present, then, under the heading of nondemonstrative arguments let us count those that are inductive (for example, 'Caesar, Fido, Jock, and the big Borzoi each bark; so probably all dogs bark') and also others which are not obviously inductive (for example, 'The orbit of Uranus shows fluctuations such as might be caused by the gravitational influence of a more remote planet; therefore, Neptune probably exists').

It is sometimes said that the chief difference between nondemonstrative arguments and demonstrative ones is that arguments of the former kind yield new information—their conclusions are not "contained in" the premises; whereas arguments of the latter kind merely tell us what we already know, merely unfold conclusions that were already "contained in" the premises. This way of speaking is wholly unilluminating, however, for it sheds no light upon the logic of the situation. That nondemonstrative conclusions often strike us as being novel is merely a psychological fact (if it is a fact); moreover, in a complicated deductive argument the demonstrative conclusion may look very novel too. And to say that the conclusion of a demonstrative argument is "contained in" its premises is to use a spatial metaphor which really can mean nothing more than that the conclusion does follow from the premises demonstratively.

4

Introductory

Returning to the question of the difference between foolish, unreasonable beliefs and reasonable, sensible ones: we can say in regard to belief in empirical statements that such belief is reasonable if and only if the statement believed either can be conclusively verified by appeal to experience or (what is more interesting) is well confirmed by the evidence that experience provides. But this leads us to a more specific question. What rules determine the conditions under which evidence may be said to confirm hypotheses; that is, to what general canons of logical validity are nondemonstrative arguments subject? It is this question with which the following chapters aim to deal. First, however, let us consider the importance of investigating it, and then we can proceed to inquire whether the question may be expected to admit of any answer.

II

But is there really any point in investigating the logic of induction (or of nondemonstrative inference generally)? Are not its principles already quite clear enough? This is a proper point to raise; and certainly it is true that a number of acute writers have busied themselves with attempts to make clear the fundamental logical principles underlying inductive argument. But despite these efforts, the student of inductive logic scarcely can help feeling some qualms about the subject when he contrasts its state with that of deductive logic. The latter at the hands of its best expositors has attained a degree of rigor in its codification and of clarity about its principles that contrast rather sharply with the inconclusiveness, the ambiguities, and the disagreements of principle which seem still to affect inductive logic.

This contrast is heightened by the practice many textbook writers have of composing books in which deductive and inductive logic are coupled, as two co-ordinate branches of the same stem. The student may be disappointed then to find that the second branch offers only a job lot of rules of thumb, lacking systematic organization. Moreover, he may notice not only that the various authorities are not in agreement about what basic principles to employ, but also that many of them do not even notice these disagreements, and still less do they present convincing arguments pro and con concerning the relative merits of the principles which they variously espouse. Of course one cannot expect too much from textbooks; but even substantial treatises on inductive logic often do not succeed in presenting a view of the field as a whole. The various parts of the subject frequently are not brought into connection with each other; too often no notice is taken of related epistemological problems, and discussion is pursued in an artificial atmosphere which precludes any very satisfactory exposition of first principles.

Thus it does seem fair to say that further investigation of inductive logic really is called for; investigation might profitably be aimed at better understanding of its basic principles. Such investigation may not prove very successful, but surely it is worth attempting. And furthermore, even if inductive logic were not of interest in its own right, it would merit investigation on account of its intimate connection with philosophical problems in metaphysics and in epistemology. It is desirable that inductive logic not be studied in abstraction from philosophy; but it is equally desirable that philosophy not be studied in complete abstraction from this logic.

One philosophical problem to which inductive logic is relevant is that of other minds. Roughly expressed, the problem is this: what right have I to regard other people as more than mere automata? I can observe their overt behavior, but I never directly observe any sensations or thoughts or feelings of theirs. Have I any right, then, to believe that they are conscious at all? Does what I observe of the behavior of others provide evidence confirming the hypothesis that they are conscious beings; or is that hypothesis an unconfirmable, nonsensical one? If it were logically impossible to use evidence about outward behavior in order to confirm hypotheses about inner mental states, then perhaps we ought to embrace the behavioristic view that speaking of someone else's mental states is just a covert way of describing his behavior. Clearly this problem is of considerable philosophical interest; and it can hardly be discussed without appeal to inductive logic.

Another philosophical problem to which inductive logic is relevant is the perennial problem of realism. On the one hand, philosophers who are realists have argued that there is an external world in which physical objects exist whether they are experienced or not; on the other hand, subjectivists, phenomenalists, and idealists have continually argued that this hypothesis is absurd and that only things which are experienced can be said to exist. This long controversy, often so inconclusively pursued, springs out of one central issue: whether any argument can be cogent which takes as its premise evidence from experience and proceeds to the conclusion that there probably exist entities which are not experienced. The realistic doctrine is regarded as an empirical hypothesis by its more sensible proponents, who realize that it could not be established a priori. They suggest that this hypothesis is confirmed by the fact that sense-experience proceeds just as if it were true and that the hypothesis is a legitimate one because it has a real explanatory value. It would be wildly improbable, they contend, that the observed regularities among sense-experiences should occur on the basis of chance alone; and this, it is said, confirms the realistic hypothesis. The realist, if he is to make his case convincing, must rely upon applications such as these of nondemonstrative arguments.

Subjectivists, however, will reply that any such argument as this involves an illegitimate kind of nondemonstrative reasoning. They will insist that if there were objects which existed unobserved in an external world then we could not have any evidence of their existence. From the evidence that such-and-such experiences occur one would be entitled to infer the conclusion that there exist unexperienced objects corresponding to these experiences only if one had already established that there generally do exist unexperienced objects corresponding to experiences of these sorts. But the latter generalization is something that experience never could establish, since one cannot experience an unexperienced object. The subjectivist will conclude that we therefore cannot possibly have any evidence of the existence of unexperienced objects; no argument can be valid which purports to obtain the conclusion that independent things exist from the premise that sense-experiences of such-and-such sorts occur. The realistic hypothesis is not a legitimate empirical hypothesis at all, the subjectivist will say, because no possible evidence could confirm or disconfirm it; and it has no explanatory value, because it implies nothing about the content of experience.

It is considerations like these, pro and con, which have formed the staple of the controversy concerning the realistic hypothesis. And these considerations clearly have a direct connection with the logic of nondemonstrative inference. Is it the case that the realistic hypothesis is a legitimate one, or is it impossible that any evidence ever could confirm it? The matter is fundamental to much of epistemology and metaphysics. Yet it is difficult to answer such questions intelligently, and no hasty or doctrinaire answer can carry weight. To attempt answers without first undertaking a thorough investigation of nondemonstrative logic at best is idle, at worst may be very misleading.

8

Introductory

Finally, nondemonstrative logic also is connected with still other, rather more concrete problems. Even if it be thought odd for anyone to brood over such an esoteric question as whether the external world exists, there remain other perplexing questions to which this logic is relevant, questions of more common interest, questions that can quite naturally arise in the mind of the ordinary man when he reflects about the theories of modern science. Are there really such things as quanta and electromagnetic waves? Do electrons exist? Are needs and drives, libidos and superegos real? Or are scientists actually indulging in elaborate *façons de parler*, or fictions, or self-deceptions, when they talk of these things as though they existed? One's interest in these questions is likely to be stimulated by the somewhat shocking readiness with which many scientists, when for a moment they turn philosophical, are willing to disclaim any concern with reality. Such questions as these are philosophical in that they call upon us to re-examine the criteria that ought to be employed in distinguishing between real and unreal; yet they are questions concrete enough to be of interest to almost any thoughtful person; and they are questions which it would be strange to dismiss as senseless. They are questions which in some cases may even have a practical bearing upon the actual development of science itself. The philosophical view that scientists take as to the status of the entities mentioned in their theories is likely to influence what they are willing to say about them, and at some critical junctures it may influence the scientist's choice of scientific theories. The scientist might actually get into trouble as a scientist if he were too neglectful and careless in his philosophical opinions. The tradition of good sense which governs the work of those in the exact sciences usually precludes this; but in the social sciences the danger is commonplace. If the danger is to be avoided, questions like these need to be dealt with explicitly; but they cannot be dealt with apart from inductive logic. To answer them we must consider what sorts of scientific conclusions admit of being established through use of nondemonstrative inference.

A variety of considerations thus combine in enjoining us to seek a more thorough understanding of the logic of nondemonstrative inference. We need to know whether there are principles which ought to govern such inferences; if there are, we need to know what they are, and we need to know what sort of justification can be offered in their favor.

III

In the preceding section we concluded vaguely that the logical basis of nondemonstrative inference needs investigating. But of what sort of investigation does it admit? In what sense is it possible to criticize and to assess the validity of different modes of nondemonstrative inference? It is necessary to consider this, because many contemporary philosophers seem to hold that there cannot be any genuine philosophical problems concerning such inferences. They seem to hold that it would be senseless to attempt philosophically to justify or to criticize nondemonstrative arguments. This is a serious charge, and to proceed without having examined it would be to invite misunderstandings.

Actually, most of the writers who have dealt with nondemonstrative inference have laid especial emphasis upon what is called "the problem of induction." This is Hume's problem: what right do we have to suppose that the future will be like the past? Or better, what right have we to suppose that certain information about what has been observed can confirm certain hypotheses about what has not been observed? It seems widely to be supposed that there are no philosophical problems about nondemonstrative argument which do not reduce to Hume's problem; moreover, nowadays many philosophers have concluded that Hume's problem is not a genuine philosophical problem but only a confusion. In earlier decades, Hume's problem used to be taken more seriously; Russell and Keynes and others felt sure that it raises genuine and important philosophical issues. Nowadays, however, a commoner view is that "the problem of induction" is something deserving to be dissolved rather than to be solved.

We started off by trying in an introductory way to formulate some philosophical questions about nondemonstrative argument. Were we drifting into nonsense? Does induction (and nondemonstrative inference generally) give rise only to pseudo-problems, to linguistic confusions? Or are there really some issues involved here that do call for philosophical investigation, some questions that do merit answers? Let us try to see what some of the considerations are that have been advanced by those who claim that no genuine problem is involved.

Let us speak first of one obvious yet important point of view according to which no genuine philosophical problem about the logic of nondemonstrative inference could arise: this is the view that we do *not* have any right to regard nondemonstrative reasoning as trustworthy, that no nondemonstrative argument is any better than any other, for all are uniformly invalid. This essentially was the view of Hume himself, during his characteristic skeptical moments. And more recently this point of view has received classic expression in Wittgenstein's *Tractatus*:

The process of induction is the process of assuming the *simplest* law that can be made to harmonize with our experience.

This process, however, has no logical foundation but only a psychological one.

It is clear that there are no grounds for believing that the simplest course of events will really happen.

Induction and Hypothesis

That the sun will rise to-morrow, is an hypothesis; and that means that we do not *know* whether it will rise.[1]

Here one can see that the conclusive arguments of deductive logic are being taken as the paradigm; and nondemonstrative arguments, because they are not conclusive, are rejected as logically worthless.

This point of view has itself a certain attractive simplicity; but it has also some rather distressing consequences. Suppose that I have a practical problem to decide; for instance, perhaps I have to decide whether to build my house near a river which sometimes overflows. If I regard as highly probable the hypothesis that this river will not flood the building site during the next twenty years, then I shall decide to build there, otherwise not. Suppose that two persons present themselves to assist me: one is an engineer who has carefully studied the past performances of the river and who assures me that this is a dangerous place on which to build; the other is an old gypsy woman who, her palm having been crossed with silver, consults a crystal ball and then assures me that the site is perfectly safe. Here we have two quite different methods of proceeding, and they yield different predictions: could anyone seriously maintain that these methods are equally irrational? Surely not. No doubt it is a psychological fact that we in our culture do happen to pay more attention to scientists than to gypsies; but that is not all there is to it. There is a further important fact: we are *right* to do so, for scientific methods somehow are more logical, more rational, than are gypsy methods. Only someone who was thoroughly disingenuous (or perhaps thoroughly confused) would actually go about advocating the hiring of engineers rather than of gypsies yet at the same time maintaining that no method of appraising hypotheses is any more reasonable than

[1] Ludwig Wittgenstein, *Tractatus Logico-Philosophicus* (London, 1922), 6.363.

any other. To advocate a thoroughgoing skepticism here, to maintain that all methods of appraising hypotheses are equally irrational, would be to embrace a stultifying philosophy, which no one but a vegetable could consistently maintain. As active beings all of us are continually obliged to make decisions in the light of empirical hypotheses; and we are obliged to believe that some hypotheses are better confirmed than others. It would seem, therefore, that we must set aside this skeptical first objection.

Let us speak next of a second influential objection that has been raised against the legitimacy of any philosophical problem about nondemonstrative argument. A number of philosophers who take "the problem of induction" to be central have construed this as being the problem of finding a conclusive proof or guarantee that the use of certain inductive procedures must necessarily be successful in leading to true conclusions, at least "in the long run." [2] Presumably, in order to show that the engineer's method of reasoning is rational and the gypsy's method irrational, one would have to prove that the engineer's method, if pursued long enough, would necessarily lead to the truth and that someone who persisted in using the gypsy's method might never find out the truth about the matter under investigation. The problem of induction having been formulated in this way, these philosophers go on to observe that, in the nature of the case, no such conclusive proof could possibly be constructed; it is impossible to prove that some particular inductive method must necessarily lead us to true conclusions. The problem of induction, therefore, is self-contradictory and must be dismissed, these philosophers say. We had better just stop brooding about this pseudo-problem, and indeed about the whole subject.

---

[2] For example, G. H. von Wright, A *Treatise on Induction and Probability* (London, 1951).

Induction and Hypothesis

Now, there is an element of truth here. If our problem were that of finding a conclusive proof that use of certain inductive procedures must lead to true conclusions, at least "in the long run," then we should indeed be deluding ourselves with a false problem. This reflection, however, is not new; more than seventy years ago Bradley wrote: "But what is this 'long run'? It is an ambiguity or else a fiction. Does it mean a finite time? Then the assertion is *false*. Does it mean a time which has no end, an infinite time? Then the assertion is *nonsense*." And again, "We do *not* know that in the long run the events will correspond to the probabilities. . . . It is mere superstition which leads us to believe in the reality of the fiction which gives rise to these chimaeras." [3] Bradley saw through this false problem. And of course it is important for us to realize that there cannot be any demonstrative proof that any specific sort of nondemonstrative reasoning must lead to true conclusions; but surely this could be accepted as a commonplace today. The better writers about induction have not imagined that it was their duty to establish any such impossible thesis. The worth-while questions about induction remain: what mode of nondemonstrative argument should be regarded as the fundamental valid one; why ought it to be trusted in preference to other possible modes of argument? Why trust the engineer more than the gypsy? What we require is some sort of philosophical explanation of why one method is to be regarded as more reasonable than other alternative methods. There seems to be no reason for regarding these as self-contradictory or absurd questions.

So far, two objections have been mentioned which purported to show that any philosophical problem about nondemonstrative argument was nonsense; and neither of these objections seemed convincing. Let us consider next some further objections which

---

[3] F. H. Bradley, *The Principles of Logic*, 2d ed. (London, 1922), I, 288, 292.

purport to show that questions about induction (or about non-demonstrative argument generally), while perhaps not exactly nonsensical, at any rate are nothing that we ought to be perplexed about philosophically, because they admit of easy answers.

IV

Some people hold that the only general question that it is sensible to ask about an inductive method is the empirical question, does it actually work? If you are in doubt about whether the engineer's method of making predictions is better than that of the gypsy, then just look and see which of them has had the better record of success so far. The inductive methods that engineers and scientists use have worked, they usually have led to conclusions that were true, or at any rate useful; whereas predictions made by gypsies have often turned out to be mistaken. Since scientists have been more successful so far than have gypsies, in all probability scientists will be more successful in the future also—and this gives us good reason for preferring scientists to gypsies. The justification of scientific method is that it works. Pragmatists are especially addicted to this sort of view; following Dewey, Nagel for instance speaks of the principles of scientific inference as "warranted by their matter-of-fact success" in "identifiable contexts of inquiry." [4] There is no philosophical problem left over; the progress of science proves the soundness of whatever methods of nondemonstrative reasoning scientists do use, and that is all there is to it. The hyperbolic perplexities about induction, in which some philosophers indulge, are pointless and uncalled for. This account of the matter is down-to-earth and may appear to bear the stamp of wholesome common sense. Unfortunately it is circular.

[4] Ernest Nagel, *Sovereign Reason* (Glencoe, Ill., 1954), *passim*.

Induction and Hypothesis

The circularity involved becomes evident as soon as we ask, how are we to find out whether the modes of nondemonstrative reasoning that scientists employ do lead to results that are true (or at any rate useful)? Clearly, we must conduct a scientific inquiry in order to discover whether the people who employ these methods actually do obtain true or even useful results through their application. But this inquiry which we are to conduct will itself have to be an empirical inquiry, it will have to employ some mode of nondemonstrative reasoning, and the conclusion we reach can be no more trustworthy than is the method we use in order to reach it. Thus only if we are willing to presuppose the trustworthiness of the modes of scientific inference that we employ, can we repose any confidence in the information obtained through their use—the information, that is to say, about whether these very modes of inference themselves are "warranted by their matter-of-fact success." If one assumes a certain kind of nondemonstrative reasoning to be sound, one can perhaps show empirically that this kind of reasoning usually leads to true or useful results. But as a proof of the trustworthiness of the method, this is circular and begs the question. As it stands, such an argument gives us no reason whatever for trusting the particular method of nondemonstrative inference, whatever method it may be, that we choose to employ.

The inadequacy of this attempt to use nondemonstrative reasoning in order to confirm the trustworthiness of a method of nondemonstrative reasoning becomes especially clear when we reflect that many quite different rules of nondemonstrative inference can be "self-confirming." To say that a rule is self-confirming is to say that the hypothesis that this rule leads to true conclusions is a hypothesis which can be confirmed through use of the rule itself. Burks has constructed some elegant examples of this sort;[5] but for simple example we may recur to our gypsy woman. Suppose that her customers restively inquire whether her method of appealing to a crystal ball in order to establish hypotheses is a method which actually works. The gypsy may reply, "Oh, yes indeed, you may be sure that my crystal gazing yields reliable answers to all your questions. I know that it does, for I conducted an empirical inquiry into the matter; seeking an answer to the question whether my crystal gazing is a reliable way of answering questions, I looked into my crystal ball, and the answer that I saw there was 'Yes.' Thus I have proved the reliability of my method; I have proved that my method is warranted by its matter-of-fact success in identifiable contexts of inquiry."

If the gypsy were to reply in this way, pragmatists would have to accept her reply, it would seem. But this we surely must regard as a *reductio ad absurdum* of the pragmatic point of view here. Innumerable quite different inductive methods can be self-confirming; hence the mere fact that it is self-confirming cannot be regarded as a significant argument in favor of any particular method. Our problem of justifying one method as opposed to others remains a philosophical problem, not a problem to be settled by ordinary empirical investigations. Indeed, one might go so far as to say that, when we are faced with a choice among different fundamental modes of inductive inference, all considerations of practical success are irrelevant. Just as the moral virtue of the motives upon which one acts does not depend upon whether one happens actually to be successful in producing good results, so the rationality of one's beliefs in no way depends upon whether or not they happen to

---

[5] A. W. Burks, "The Presupposition Theory of Induction," *Philosophy of Science*, XX (1953), 177–197. See also Max Black, *Problems of Analysis* (Ithaca, N.Y., 1954), ch. x.

be true. If the sun does not rise tomorrow, if Queen Anne still lives, that does not prove that our beliefs to the contrary were irrational. There might even be a world—under the governance of a malicious Cartesian demon perhaps—in which virtuous actions always led to bad results and rational beliefs always were false: yet even in such an undesirable world virtue would still be virtue and rationality, rationality.

V

In the preceding section we considered the view that the only general question that it is sensible to ask about various inductive methods is the empirical question, do they work? We saw that this point of view is unacceptable. There remain, however, some other objections which purport to show in other ways that no philosophical problem about induction (or about nondemonstrative inference generally) can arise.

One of these objections, likely to be offered by Logical Positivists, centers about the notion of validity itself. Logical Positivists hold (or at any rate, in their salad days they used to hold) that any significant assertion must either be an empirical statement or else must be an arbitrary stipulation about how words are going to be used. Now, consider the assertion that such-and-such a mode of nondemonstrative inference is valid, while other modes are not valid. What is the status of an assertion such as this? We have already seen that it cannot satisfactorily be regarded as an empirical statement; therefore, the positivist would argue, it must be regarded as nothing but a stipulation or proposal regarding the use of the world 'valid.' The question whether a specific mode of inductive reasoning is valid is just a question about how words are to be used. But in that case, it is no very serious question, since words may be used in accord with whatever conventions we please. Usually it is desirable to follow the linguistic precedents set by the scientists of our culture circle; therefore, let us solve the problem of induction simply by deciding that we shall apply the word 'valid' to whatever methods of reasoning scientists do employ.

The proponent of this view doubtless conceives it to be the essence of sweet reasonableness. But beneath this veneer lies irrationalism. For according to this view, any choice among various alternative methods of nondemonstrative reasoning must be wholly arbitrary and dogmatic. Not only is it denied that we do have any reasons for trusting scientists more than we trust old gypsy women; it is denied that there could be any such reasons. Why then should we adopt the linguistic conventions of scientists rather than those of gypsies, or astrologers, or mystics? Why not call the gypsy's method of crystal gazing "valid" and its conclusions "well confirmed"? People holding this view, if they really were tolerant, would abandon their blind prejudice in favor of scientists and would give the gypsies a fair deal.

Surely we cannot but conclude that this objection is ill-taken. It is intolerable to say that there is no reason why science is any more trustworthy than crystal gazing; none of us for an instant seriously believes this. Blind dogmatism and utter skepticism are equally stultifying, and no sensible person could seriously embrace either one as an adequate account of the basis of empirical knowledge.

Let us pass to another, more sophisticated sort of objection, which is offered by Wittgenstein in the *Philosophical Investigations*. He writes:

But how *can* previous experience be a ground for assuming that such-and-such will occur later on? —the answer is: What general concept have we of grounds for this kind of assumption? This sort of statement about the past is simply what we call a ground for

assuming that this will happen in the future. —And if you are sur-
prised at our playing such a game I refer you to the *effect* of a past
experience (to the fact that a burnt child fears the fire).

A good ground is one that looks *like this*.[6]

Now, these remarks no doubt are profound, and one would
hate simply to be guilty of misunderstanding them; but it does
seem as though they give a misleading impression. It seems that
Wittgenstein is here suggesting that a bit of sensitive reflection
upon ordinary usage would suffice to resolve anything that purported to be a philosophical problem about induction. And
surely this is not really so. Surely there are philosophical questions about induction which the most sensitive lexicographer
would be at a loss to answer. The use of the phrase 'good
grounds' as we meet it in ordinary discourse is not very clear
in some respects; and it would hardly seem that any direct
appeal to the ordinary use of this phrase could succeed in
providing answers to some of the questions that we need to
have answered.

The considerations suggested in section II underline this.
Those who write about inductive logic do not manage to agree
concerning its principles; philosophers who argue about other
minds or about the external world usually do not seem to be
clear about just what modes of nondemonstrative argument
ought to be considered valid; and even scientists in concrete
cases sometimes appear unable to decide definitely what kinds
of hypotheses their evidence does support. Now, when people
disagree about the validity of an argument, surely we should
want to say that one party is in the right and the other is in
the wrong (to say that given evidence neither does or does not
provide good grounds for believing a given hypothesis surely
would be to misuse the phrase 'good grounds'). But it seems

[6] Wittgenstein, *Philosophical Investigations* (Oxford, 1953), pt. 1, 480,
483.

Introductory

impossible to see how the dictum "A good ground is one that looks *like this*" can enable us to decide which is right. It seems impossible that such arguments should be resolvable merely by direct appeal to the ordinary use of language, for ordinary people do not agree (nor do scientists) in their use of the phrase 'good grounds' or the world 'valid.'

There remains one further, allied objection that ought perhaps to be considered. This is the objection that inductive inference (and presumably nondemonstrative argument generally) does not admit of any formal rules, that it does not have any real principles, and consequently that there can be no philosophical questions to raise concerning the nature or trustworthiness of its principles: it does not have any. Some such view as this was propounded by the Absolute Idealists, but has been embraced by others more recently.[7] Those who hold this view claim that the validity of any specific inference can be judged only by considering that specific case; each case is *sui generis*, a law unto itself. General, formal rules about validity are, at best, only hypotheses which must be tested by seeing whether they hold in specific cases; and if an ostensibly general rule seems to disagree with a specific case, it is the ostensibly general rule, not the specific argument, which must be regarded as incorrect. Moreover, any general formal rules which we may construct, though they will perhaps fit many cases, will never succeed in holding in all cases; specific arguments in their concrete richness always overflow the Procrustean forms within which we seek to confine them. Nondemonstrative inference is an art, not a science, for it must proceed by means of insight into the specific rather than by means of abstract general rules.

There are several things that need to be said in reply to this objection. To begin with, it is likely that those who propound

---

[7] For example, P. F. Strawson, *Introduction to Logical Theory* (London, 1952), p. 248.

Induction and Hypothesis

it are suffering from the old confusion between the inventing of arguments and the evaluating of them. The inventing of good arguments may indeed be an art; at any rate, it does not readily admit of being reduced to a mechanical procedure, but rather calls for inspiration, intuition, and creative ingenuity. However, the business of deciding whether an already formulated argument is valid or invalid is quite a different matter; this decision must admit of being made in accord with general rules, it cannot just be a matter of happy intuitions. For suppose that there were no rules governing such decisions: what would happen when people disagreed in their intuitions concerning the validity of some given argument? In such a case either some persons' intuitions would be correct and conflicting intuitions incorrect, or else no intuition is any better than any other. Now, if one person's intuitions are correct, then there seems to be no reason why in principle some general rule should not be stated which would express the difference between valid and invalid arguments. For if a human faculty of intuition (we need not inquire into the psychological status of such a supposed faculty) can effectively discriminate between valid and invalid inferences, then a machine could in principle do the same job (any such task that can be done, can be done by a machine); and of course a machine will be designed so as to operate according to general rules. On the other hand, if no intuition is any better than any other, we are plunged back into stultifying skepticism, which cannot be a serious view.

Indeed, if an argument is to be of any use, its validity must admit of being grasped by people who may not be capable of such difficult fits of inspiration as may have gone into its construction. A genius may be able to see at a glance that the conclusion in a complex argument really is supported by the premises; but surely the essence of an argument is that it should exhibit step by step (with a rule ready for citation at each step)

22

just how the conclusion is connected with its premises, so that the diligent but less gifted person may also understand.

Furthermore, the proponents of this view make a mistake, surely, if they suppose that, just by inspecting a specific argument, one can always reach a clear decision as to whether it is valid or not. In a preceding section we noticed some arguments (concerning the existence of unobserved entities) which arise in science and in philosophy; common sense yields no clear-cut answer to the question whether these arguments are valid. When we reflect carefully about one of these specific arguments, our intuitions sometimes balk and fail; the "natural light" gutters and grows dim. If we are to reach any sort of sensible decision concerning the validity of difficult arguments such as these, we can do so only by seeking general principles which they can be brought under. Intuition undisciplined by rules is bound to be beset by confusions and obscurities; whereas it is quite possible that the correctness of a general rule, which in being general is purged of irrelevant content, may sometimes be more evident and easier to grasp than the correctness of a specific argument, contaminated as it is by distracting matter not germane to its validity. And even if it were the case that valid nondemonstrative arguments do not conform to any rules simple enough for us to be able to formulate, this could be discovered only after a thorough attempt had been made to formulate such rules—an attempt which their opponent characteristically refuses to undertake.

VI

The first part of this chapter attempted informally to pose a philosophical problem about the logic of nondemonstrative reasoning. A number of philosophers nowadays seem to hold that there are no real philosophical problems connected with induction, however; consequently some of the characteristic objections that they advance have been considered. The earlier ones among these purported to show that no sensible problem about induction could even be stated; later objections purported to show that the problem was susceptible of some rather easy kind of solution, either through empirical investigation or by attention to verbal considerations; and the final objection considered was that there could be no problem here because inductive reasoning does not follow rules. Each of these objections, upon examination, turned out not to be convincing. It would seem that we may be entitled to conclude that there is a legitimate philosophical problem about the nature of nondemonstrative inference, a problem to which solution would be more appropriate than dissolution.

What sort of problem is this, and what kind of considerations could be relevant to its solution? We can at least say that it is not an empirical problem and no empirical evidence can be directly relevant to it. It must be a philosophical problem, in that it calls for examination of the basic logical principles which are fundamental to empirical knowledge. We need to reflect as best we can upon the fundamentals of nondemonstrative reasoning, and we need to try to develop a general theory about its logic, a theory which can systematize our intuitive beliefs about what sorts of reasoning are valid and which can at the same time possess some intrinsic plausibility of its own. Such a theory would help us to make up our minds concerning the perplexing cases which we are at a loss whether to classify as valid or as invalid. In a way, of course, this involves a Kantian approach to the matter: that is, we start with the supposition that empirical knowledge is possible, it being absurd to deny this; but we seek then to know how it is possible—we seek to know by what mode of nondemonstrative reasoning empirical knowledge may validly be built up. There must be some mode

Introductory

of nondemonstrative reasoning that is valid, and we seek reasonable grounds for deciding what it is.

The methodological remarks are vague; but it would be wrong, surely, to insist that before starting to answer a question one had always to address oneself first to another question, by what method shall my original question be dealt with? What is important is to try to answer the original question itself. And the natural procedure to adopt in discussing nondemonstrative inference is to consider the principal methods of inference that have actually been proposed and defended. By eliciting some of the merits and defects of each, we may be advanced toward an understanding of what logical principle ought to be regarded as fundamental to nondemonstrative inference and toward a comprehension of the reasons that can be offered in its favor.

$\mathcal{T}wo$

Hypotheses and Evidence

I

INSOFAR as it goes beyond the evidence upon which it is based, all our empirical knowledge about the world may be said to consist of conjectures, or hypotheses. Of course it is not enough merely to be supplied with a mass of uncriticized hypotheses, however; only insofar as we are able to evaluate these hypotheses do we possess knowledge. Now, there are philosophers who say that the task of evaluating hypotheses consists merely in rejecting some and accepting the remainder.[1] But this schematization surely is somewhat misleading, for in everyday life as in science we must not forget that hypotheses possess various degrees of credibility. We should hardly care to entrust our defense to a general who when confronted with conflicting hypotheses about the disposition of the enemy made it his practice simply to assume the more credible ones to be true and

---

[1] For example, R. B. Braithwaite, *Scientific Explanation* (Cambridge, 1953), p. 355.

26

the remainder to be false; such a careless practice would lead to fatal neglect of many contingencies for which it might have been possible to prepare. We should admit that on the basis of the evidence actually available some hypotheses will be quite credible, some will be quite incredible, but others will enjoy intermediate degrees of credibility. It could be said that to possess empirical knowledge is (at least in part) to be able correctly to assign to various hypotheses the degrees of credibility appropriate to each on the basis of the actual empirical evidence. This assignment cannot be an arbitrary one; there must be a correct way of making it, in accord with some general rule—for otherwise we should not be entitled to claim that we possess empirical knowledge.

It must be that empirical evidence of the available kind can, according to some logical rule, confer upon various empirical hypotheses various appropriate degrees of rational credibility. Thus three factors are involved: there must be evidence, there must be hypotheses, and there must be a logical principle. The logical principle no doubt is the most crucial member of this trinity; but before we try to discuss it directly, it will be helpful first to consider the nature of the hypotheses and of the evidence.

II

What logical forms may empirical hypotheses have? Must they be restricted to certain special forms if they are to be confirmable?

Since the time of Aristotle there has been a philosophical tradition to the effect that science treats of universal generalizations (statements of the form 'All ravens are black'). This has sometimes been taken as the distinguishing characteristic that marks off science from other sorts of empirical knowledge. Science seeks to establish universal laws, it is said, whereas history and other inferior departments of knowledge deal only with hypotheses particular in form ('Caesar crossed the Rubicon'). Whether or not this is accurate, universal generalizations are at any rate one of the important kinds of hypotheses. But clearly we cannot restrict ourselves to hypotheses of this universal form, for both in everyday life and in science it often is important to evaluate hypotheses of other forms—hypotheses about particular facts or events being the most obvious examples ('Life on earth began two billion years ago,' 'This galvanometer is in working order'). Actually it would seem wisest for us to allow that empirical hypotheses may be couched in any of the innumerable truth-functional or quantificational forms countenanced by deductive logic; we want to be prepared at least to consider hypotheses of any of these forms. We do not want a theory of confirmation that will apply only to universal hypotheses, or only to particular hypotheses, or only to hypotheses of any other such limited form. We want a theory which will be able to account for whatever degree of confirmation hypotheses of any truth-functional or quantificational form may come to possess.

However, is the ordinary logic of truth-functions and quantification sufficient to enable us to express our empirical hypotheses? Does something further, some notion of causal necessity, need to be added to our logic?

Consider the scientific law, 'All bodies that fall freely at sea level accelerate at the rate of 32 feet per second per second.' This is a well-confirmed hypothesis, and one of an important kind. From the point of view of ordinary logic, we should be inclined to symbolize this statement by means of the horseshoe, taking it to be of the form '(x) $(Fx \supset Gx)$'; that is, we should take it to mean that there are no freely falling bodies at sea level which do not accelerate in this way. But some philosophers would say that this interpretation of its logical form is incorrect;

they would say that the statement about freely falling bodies does not merely assert that there are no freely falling bodies which do not accelerate in this way (this assertion might be true simply in virtue of the fact that no bodies do fall absolutely freely). They would claim that the statement makes a stronger assertion, namely that if anything *were* a freely falling body, then it *would* accelerate in this way. They would claim that the statement asserts that a thing's being a freely falling body at sea level causally necessitates its accelerating in this way. Ordinary logic has no symbol for expressing the notion of causal necessity supposedly involved in this stronger assertion; if we want to express it, we must introduced a new symbol.[2] According to this view, scientific laws (the most important hypotheses to be met with in science) always do involve a notion of causal necessity—and this is best brought out by translating the laws into subjunctive conditional statements.

It is unnecessary here to enter into any protracted discussion of the problem of subjunctive conditional statements.[3] But three points perhaps deserve mention. In the first place, subjunctive conditionals are nebulous statements, and it is easy to lapse into nonsense if one uses them carelessly. The sense of a subjunctive conditional statement is very heavily dependent upon the context in which it is used, and often even the context does not indicate at all definitely what sense the statement has. In such cases it may be quite impossible to determine what it would mean for the statement to be true or to be false (compare 'If G.M. were manufacturer of the M.G., then the M.G. would be a big car' and 'If G.M. were manufacturer of the M.G., then G.M. would be manufacturer of a small car'; does

---

[2] A. W. Burks, "The Logic of Causal Propositions," *Mind*, LX (1951), 363–382.

[3] Nelson Goodman, "The Problem of Counterfactual Conditionals," *Journal of Philosophy*, XLIV (1947), 113–128.

Induction and Hypothesis

it make sense to ask which of these two statements, just as they stand, is true?). When we wish to speak exactly, we should be wise to avoid subjunctive conditionals; indicative statements are less troublesome.

In the second place, in the more advanced sciences subjunctive conditional statements do not occur frequently. In a treatise on mechanics such a statement as 'If a body were to fall freely at sea level, then its acceleration would be 32 feet per second per second' is not likely to occur at all; in its stead there will be more general statements of ordinary indicative forms, statements about the ways in which actual bodies do behave. We are entitled at least to the pious hope that it may be possible to state any law needed for science without employing subjunctive conditionals or any notion of causal necessity.

In the third place, we shall have a difficult enough time as it is, trying to explain how a statement of the form '(x) $(Fx \supset Gx)$' can be confirmed by empirical evidence. To try to show that statements of a form stronger than this (statements involving causal necessity) can also be confirmed would be a painfully difficult task. We never can verify by direct observation that one particular phenomenon causally necessitates another particular one; how, then, can we hope to infer that every phenomenon of a certain kind causally necessitates one of another kind? If the evidence can support the hypothesis that all F's are G's, we ought to try to rest content with this and not demand confirmation of the far stronger hypothesis that being F "causally necessitates" being G. Indeed, the latter hypothesis appears to be more a metaphysical than an empirical hypothesis; for it seems quite impossible to say what evidence, over and above the evidence relevant to confirming '(x) $(Fx \supset Gx)$' could be relevant to its confirmation. Therefore, it would seem wiser not to insist that scientific hypotheses need involve any notion of causal necessity.

30

Hypotheses and Evidence

Empirical knowledge may be regarded as a fabric of hypotheses, hypotheses confirmed by the evidence which experience provides. But what is the nature of this evidence? What character must the statements have in which this evidence is expressed? To be sure, one might wish to work out a theory of confirmation which imposed no restriction at all upon the character of the evidence statements, a theory which would determine the degree of confirmation that any hypothesis would have with respect to any body of statements whatever considered as evidence; this might be interesting for its abstract generality. However, from a philosophical viewpoint, it is desirable to consider whether statements only of some special kind really deserve to function as evidence—for if this is so, we shall want to focus our attention upon such statements and upon their connection with hypotheses.

The evidence upon which empirical hypotheses can be based must itself be empirical evidence; the statements in which it is expressed must be statements whose truth can somehow be certified by appeal to experience. But there are two vexed questions here: ought the evidence statements to be known with certainty? And ought they to be statements describing the observer's immediate sense-experiences? Philosophers have been divided in their answers to these questions. Some have held that these evidence statements must be certain and therefore must be statements about immediate experience. Others hold that there is no need for them to be certain and that they ought rather to be statements describing physical states of affairs. We had better consider this matter, even if we do not decide it.

Why have some philosophers argued so insistently that empirical knowledge must ultimately be based upon evidence which is known with certainty? And why have they maintained that such evidence must be evidence about one's own immediate sense-experiences? They reason in the following way. Probability, or support, or confirmation (we need not distinguish among these words) is essentially relative. A hypothesis is confirmed or is probable not all by itself but only relative to the evidence that is actually available—if the available evidence were different, then the degree of probability of the hypothesis might be completely altered. For instance, if all the evidence we possess about Jones is that he is thirty years old and not suffering from any disease, then our common knowledge plus this evidence makes the hypothesis quite probable that Jones will be alive next year. However, if we learn that Jones has just collided with a concrete wall at ninety miles per hour, then this addition to the available evidence will sharply alter the probability of the hypothesis. The very same hypothesis possesses very different degrees of probability relative to different bodies of evidence. Now, if a hypothesis is probable relative to such-and-such evidence, and if just this evidence is known with certainty to be true, then we shall be entitled to regard the hypothesis as really probable—as something which it is reasonable to believe. But if the evidence is not certain, we cannot do this, for in that case all we know is that *if* we were entitled to believe the evidence *then* we should be entitled to believe the hypothesis. To be sure, it would suffice if our evidence itself were made probable by some more ultimate evidence we possessed which actually was certain—for in that case the ultimate evidence, being certain, could make probable the proximate evidence which in turn would make the hypothesis probable. But unless something is certain, nothing can be probable. There must somewhere be a bedrock foundation upon which we can build. Unless some basic evidence is certain, we shall have at best only a web of relative probabilities hanging in air, insufficient to establish that any hypothesis really is rationally credible. Putting it another way, suppose we had a set of statements none of which was certain but which were so interrelated that if some were true then others would probably be true: we should have so far no grounds whatever for believing any of these statements. The members of a set of statements none of which is certain cannot spontaneously confirm one another, no more than a society of beggars can raise capital by borrowing among themselves.

Philosophers who adopt this point of view are insisting that the ultimate empirical evidence upon which we base our hypotheses itself must not consist of mere hypotheses. Our evidence must be formulated in statements which are directly known to be true, statements that do not need to be confirmed by appeal to something else, statements that do not need to be justified by nondemonstrative argument. Now, what sort of empirical statements can have this character?

Ordinary statements about physical things will not do, for they cannot be altogether certain. No matter how firmly convinced one may be that a statement about some physical state of affairs is true, there always remains at least a slim possibility of error. If I seem to see before me what looks like litmus paper and if it appears to be changing from blue to red, I shall be tempted to assert, 'Some litmus paper is turning red.' But I may be deceived, for perhaps the illumination is faulty, perhaps this is not really litmus paper, perhaps I am suffering from a hallucination. Further confirmatory experiences might allay some of these misgivings. But no matter how much sense-experience one has had, one would never be entitled to say that the possibility of such error had been wholly exorcised. This shows that in making such a judgment about a physical state of affairs one is not confronting an incontrovertible delivering of sense-experience; rather one is endorsing a hypothesis, a hypothesis which in principle requires confirmation. Now, hypotheses will not do for our basic evidence; we should be arguing in a circle if we attempted to explain how hypotheses may be confirmed, yet in this explanation presupposed as evidence hypotheses themselves requiring confirmation.

The evidence must be known with certainty, and it cannot consist of statements about physical things. These philosophers will argue that there is no alternative, then, but to insist that this evidence must consist of statements about immediate sense-experience. They must be "sense-datum statements," formulated in what has been called "expressive language." [4] For immediate experiences, *esse est percipi*; immediate experiences must have whatever characteristics they are experienced as having, for that it what is meant by calling them immediate. For instance, I can be certain that there appears to me to be some litmus paper turning red. In claiming that this is how it looks to me, I am simply describing my own immediate experience, about which I can be quite sure; there is no room for error here, because I have not so far endorsed any hypothesis which could be untrue. This proposition about what I experience is the kind of thing that I can with certainty know to be true. The ultimate evidence upon which all empirical knowledge is based must be of this kind.

Some philosophers will argue in the way which we have just been considering. But there are other philosophers to whom all this talk about certainty and about immediate sense-experience is anathema. These others prefer to adopt a quite different point of view. They would contend that it is pointless to seek statements that can be known with absolute certainty. Any worthwhile statement, they would claim, is subject to the possibility of revision in the light of further evidence. Thus, in scientific procedure even reports of observations sometimes need to be revised. If an observer reports an observation that is in conflict

---

[4] C. I. Lewis, *An Analysis of Knowledge and Valuation* (La Salle, Ill., 1946), pt. II.

with well-established scientific theories, scientists perhaps will not regard this as a refutation of their theories; they may simply say that the observational report probably is untrue, that the observer probably did not really observe what he reports. From this point of view, the empirical evidence with respect to which we evaluate our hypotheses will itself consist of hypotheses— hypotheses that are fairly closely connected with observations, but any one of which may need to be revised in the light of wider experience.

Philosophers who adopt this view will be inclined to argue that it is inappropriate to picture our empirical knowledge as starting from some bedrock of certainty and then being built up step by step. Such a picture seems to them to be unrealistic and inaccurate. They would prefer to consider that in any concrete situation we always have a supply of fairly well established observational hypotheses upon which we rely, and new hypotheses are evaluated in terms of how well they fit in with this already accepted corpus. Any single observational hypothesis belonging to this body might be called into question; but we cannot attempt to question the whole body of hypotheses, for we should have no standard left in terms of which to judge the tenability of the hypotheses being doubted. The enterprise of knowledge, according to this view, is like the task of a mariner who seeks to rebuild his vessel while at sea; he can reconstruct it a piece at a time, but he cannot tear it all down and make an entirely fresh start. Or the enterprise is like that of someone who would erect a solid platform in a swamp; there is no bedrock upon which to build, but piles can be driven deeper and deeper.

Philosophers who adopt this viewpoint not only would reject the notion that empirical knowledge need be based upon evidence that is known with certainty; they also would deny the corollary that this evidence need consist of information about immediate sense-experience. They would argue that the kind of evidence that is worth while for the confirming of hypotheses must be evidence about physical things. Evidence of the sort they advocate would be formulated in "physicalistic" observational statements: these would be statements about more or less directly observable characteristics of physical objects of middling size, objects about us that can be conveniently and fairly reliably observed. Examples might be 'This meter reads "5," ' 'Some litmus paper is turning red,' 'A Geiger counter has clicked twice,' and the like. Statements like these, they would claim, express evidence of the sort that is of some practical use; it is on the basis of such evidence that we should seek to confirm other empirical hypotheses.

IV

These two conceptions of the nature of empirical evidence seem to be quite fundamentally opposed to one another. Each position has some plausibility, however, and each is able in rebuttal to adduce some philosophical arguments against the other.

Those philosophers who favor physicalistic statements would raise several objections to the view that empirical knowledge need be regarded as based upon statements about immediate experience that are known with certainty. These philosophers find repugnant the notion that we can inspect our sense-experiences and thereby derive knowledge that is certain and upon this can base all the rest of our empirical knowledge. Even if it were possible to obtain certain knowledge by inspecting one's sense-experiences, this could not possibly provide enough of a basis on which to erect the whole edifice of empirical knowledge about science and about everyday matters. Immediate experience at best would be far too fleeting and ephemeral to yield us what we require. If one tries to construct some lengthy argument in order to prove that one's present experiences confirm some remote hypothesis or other, it is likely that one's experiences, fleeting as they are, will have changed long before the argument is finished: by the time the conclusion is reached, the premises no longer will be known with certainty, for we cannot be absolutely certain about the nature of past experiences since our memory of them may be faulty. Considerations like this indicate, these philosophers would say, that the whole notion of trying to build up knowledge on an absolutely certain foundation of sense-experiences is a quixotic and utterly impracticable notion.

Moreover, these philosophers may object, the notion of private experience is unscientific. To them, talk about sense-data seems impossible, nonsensical, or at any rate pointless. The reason is that statements about immediate experience are subjective and private and do not admit of being tested in a public way. If someone says that he is having an experience of such-and-such a sort, there is no way for others to confirm or to disconfirm the correctness of what he says—if he is merely describing his own private experiences. How unsuitable it would be, these philosophers exclaim, to regard our scientific knowledge as based upon subjective, untestable statements like that. Nothing could be more unscientific, they will contend.

An allied but still stronger objection will be raised by some of these philosophers: they will say not merely that it is unscientific but that it really is nonsensical to talk about sense-data. They would claim that all our language is physical-object language, in the sense that our words refer to physical things and their characteristics. One can teach a child to use words by showing him the physical things to which the words apply or by showing him the public situations in which the words are to be used; but one cannot teach anybody a private sense-datum language for describing his experiences. Now, any language that makes sense can be taught; and a sense-datum language, being unteachable, must be senseless. Even if someone thought that he possessed such a language, this claim could not be allowed, for the very notion of a "private language" is a contradictory notion, these philosophers would say. For suppose that someone professed to have a private language in which, say, the word 'blue' was applicable to a specific sort of sensation; suppose that he claimed always to use this word correctly, applying it only to sensations of this specific kind. What could this latter claim possibly mean? Could it be true or false? No evidence could possibly count for or against the claim: the claim that he is making simply has no empirical significance whatever and makes no sense. There cannot exist any criterion for the correctness of his use of words in his private language which he professes to possess; this ostensible language therefore really is no language at all.

Criticisms of this sort could be developed at much greater length, and they may appear to be telling; yet the philosopher who advocates sense-data probably will be unmoved by them. He will admit that because of human frailty it would not actually be practicable to build up the whole of our empirical knowledge in the way he advocates; yet he would insist that it is philosophically important to examine how in principle this construction might proceed. He would argue that sense-datum statements, far from being unscientific, must be regarded as logically basic to science; of course my sense-datum statements are certain only to me, but this does not entail any unwholesome subjectivity—it merely follows from the fact that knowledge is always someone's knowledge. As for the criticism that his sense-datum language is nonsense, the sense-datum philosopher would reply that the criticism rests upon an unduly narrow assumption about what makes sense.

Mainly, however, the sense-datum philosopher will defend

himself by attacking his opponent. He will argue that if we were to adopt physicalistic statements as our evidence then we should not be able to draw any sharp and precise line between statements that are to count as evidence statements and those that are not to do so. The proponent of physicalistic statements has told us that our observational statements ought to be statements about directly observable features of physical things. But the common-sense notion of what is "directly observable" is a vague notion and yields no sharp criterion. It would seem that directly observable things are those that can be perceived with reasonable accuracy without the mediation of any unusual equipment. But this is dreadfully vague. Do we directly observe the person whose footstep we hear upon the stair? Do we directly observe the man in the iron mask? Do we directly observe through spectacles, through an opera glass, through a microscope, or through an electron microscope? Such questions as these do not admit of any definite answers, for the whole notion of direct observation is an extremely hazy one. In the nature of the case, it will be impossible to draw any sharp and definite distinction, at a physicalistic level, between what is and what is not directly observed; insofar as a distinction is so drawn as to be plausible, it will be indefinite; and to make it more definite would be to draw a line that would be implausibly artificial and arbitrary.

But if this is so, then it will become an arbitrary matter whether we regard a given statement as evidential or as hypothetical in character. And if we have no precise distinction between hypotheses and evidence, then the whole procedure of confirmation becomes arbitrary. If one class of statements is adopted as our observational statements, then one set of hypotheses will count as well confirmed; if some other different class of statements is adopted as observational statements, then some quite different set of hypotheses may have to be regarded as well confirmed instead. Contrasting and even flatly contra-dictory hypotheses may receive high degrees of confirmation on the basis of different sets of observational statements (imagine how different our theories about psychology might be if we conceded the direct observability of the phenomena which some of the more credulous psychical researchers believe themselves to have observed). If the choice of what statements are to count as observational statements is merely an arbitrary choice, then the edifice of science based upon them will be arbitrary also. But surely the practice of scientific inquiry is not merely a game which each person may play according to rules of his own taste and fancy; science is not just a tissue of fictions: it aims to get at the truth about the world. And if this is so, the principles upon which it is based cannot be arbitrary.

This argument seems to oblige us to conclude that physi-calism cannot give any convincing account of just what state-ments ought to count as observational statements. There is no clear dividing line, but only a vast region of vagueness which separates those things that fairly clearly should be called directly observable from those things that clearly ought not to be called directly observable—according to physicalism. And this under-mines the whole account of science, so the proponents of sense-data would insist.

V

Holders of these two opposed views about the nature of empirical evidence can levy dialectical objections against each other, yet it is not clear that either side is able conclusively to win the controversy. We need to pause, cultivate detachment, and reflect about the kind of philosophical issue which is at stake here. We need to reflect on what it is that a theory about confirmation, about nondemonstrative inference, is supposed to accomplish.

Hypotheses and Evidence

Now, if we were actually to go about observing people, we should find them dealing in hypotheses and constructing non-demonstrative arguments in a bewildering variety of ways, and often in ways that cannot be exactly described in logical terms. Often a person may appear to be employing an argument, yet when we look more closely we may find it impossible to determine precisely what its premises are or what the conclusion. Perhaps we know precisely what the arguer has said or written: yet if his words are vague or ambiguous it may happen that there is no way whatever of deciding precisely what the logical form of his argument is (it surely is unwarranted metaphysics to assume that "in his own mind" an arguer must always precisely "know what he means"). If we attempt to describe such an argument by saying that it has such-and-such a logical form, our description is not really accurate as a description for, instead of merely describing, what we have done is to replace the actually vague argument by a more precise logical schematization to which it perhaps bears only some affinity. Commonly when we apply logical categories to concrete arguments, this sort of replacement takes place to some extent. If the purpose of logical analysis were merely to describe the character that actual concrete arguments possess, such replacement would have to be condemned as distortion; but that is not its main purpose.

When we raise philosophical questions about the logic of confirmation, we are not really interested just in obtaining a description of what people actually do—how they act or how they use language. We are interested in obtaining an idealized scheme of argument which (whether anyone actually uses it or not) can reasonably be called valid. Such a schematization may be highly idealized and remote from actual practice, or it may be only somewhat so. What use is it? For one thing, we naturally want it to have some relevance to the reasoning in which people do indulge: we should hope that the schematization might be of some use in helping us actually to criticize concrete arguments that do occur. In virtue of having it as a standard for comparison, we should hope sometimes to be better able to judge the validity of actual arguments occurring in ordinary discourse. But there is also another additional satisfaction that we might hope to obtain from this schematization: we might hope that it would fit into our general philosophical picture in such a way as to make that picture more coherent and more complete. Those of us who are perplexed by some of the tangled philosophical issues concerning empirical knowledge will hope for clarification from this schematization.

Surely it would be harsh to say that either of these concerns is illegitimate. But they are rather different concerns, the former being of more nearly practical and the latter of more philosophical interest. Bearing this in mind, let us return to the two seemingly opposed views about the nature of empirical evidence. Perhaps their opposition will seem less extreme if we reflect that these two views respectively represent these two rather different sorts of concern.

Those who favor physicalistic observational statements are interested in developing a schematization which will not be too idealized and which will apply fairly directly to problems of the sort that present themselves to us in everyday or in scientific thinking. They seek a theory of confirmation which will assist us in criticizing ordinary arguments. And here of course they are right in maintaining that we ordinarily do employ physicalistic statements as the explicit evidence from which we start in an argument. Moreover, if incompatibilities appear, we are willing to reconsider and perhaps to abandon any particular one among these observational hypotheses that we have been accepting—though we never attempt to question all of them at once. This seems to be fairly correct as a description of actual practice; and these features seem suitable for inclusion in a logical scheme which aims to be readily applicable to ordinary thinking.

However, those philosophers have a different concern who favor sense-data and who maintain that every hypothesis we accept ought ultimately to be based upon some evidence that is known with certainty. They are concerned with trying to provide a more philosophical and necessarily more idealized schematization which will help to resolve various philosophical perplexities. They do not care much about how ordinary thinking proceeds, nor do they even care primarily about criticizing ordinary arguments. They are interested in what they take to be the presuppositions logically involved in any claim to the possession of knowledge—whether or not these turn out to be of much applicability in connection with actual practice.

No logical theory of confirmation can be a mere description of the way in which people actually argue; any theory must to some extent be an idealization which actual arguments do not entirely live up to. These two points of view differ about the degree of idealization, the extent to which the standard should be unlike the practice. Yet these two different points of view, springing from different concerns, do not really have to be regarded as opposed to one another; they might be thought of as complementary. One represents a more nearly practical point of view, the other a more abstract philosophical idealization. May we not suppose that each has its place and that neither should be lost sight of in a complete account of confirmation? At any rate, in the chapters which follow we shall leave this question open, assuming merely that in any particular case it will have been decided what observational statements are available as evidence. We shall leave open the possibility that this evidence may consist either of physicalistic statements or of sense-datum statements.

Induction and Hypothesis

Whether one decides that the evidence which experience provides is best construed as evidence about sense-data or about physical things, there still remains the separate problem of what logical form the evidence statements ought to have.

To begin with, it is clear that statements expressing our observational evidence cannot be universal in form. Direct experience cannot serve to verify any statement of the form 'All F's are G's'; for in order to verify such a statement one would have to inspect all the things that are F's and observe that each is a G, and furthermore one would have to know by observation that one was inspecting all the F's and omitting none. This last proviso (itself universal) of course could never be fulfilled; one never can be sure that further experience may not disclose further F's. That there is no unexamined F can at best be a hypothesis, requiring further confirmation; it does not admit of being verified by inspection. Consequently, any empirical universal statement must have the character merely of a hypothesis, perhaps admitting of confirmation, but never suitable for the role of evidence.

If statements which are to serve as observational evidence are not to be universal, then can they have the form of singular statements? Singular statements (containing singular terms but no universal quantifiers) perhaps do seem to admit of being verified by direct observation. For instance, if '*a*' names a particular observable thing and '*F*' is an observational predicate, then it seems that one has but to observe the thing and note the predicate is true of it, and one thereby has verified the singular statement '*Fa*.'

But before we can decide whether singular terms may occur in observational evidence, we must first consider what view is to be taken of singular terms. Singular terms include proper names, definite descriptions, and "egocentric particulars" such as 'I,' 'now,' 'here,' and the like. Suppose that with regard to all of these we were to adopt Russell's "theory of descriptions." [5] Then any statement containing a singular term would be regarded as short for some quantified statement not containing any singular term; for instance, 'Fido barks' would be short for 'One and only one thing is Fido-like and that thing barks' (to be Fido-like is to possess all those characteristics, whatever they may be, which are requisite to being Fido). But clearly if we think of singular statements in this way, we cannot regard them as verifiable by direct observation, for according to this view a universal proviso (for example, 'Only one thing is Fido-like') is involved, and such a proviso cannot be verified by observation. Consequently, if we adopt the theory of descriptions, we cannot permit a singular term to occur (unless vacuously) in any statement that is to be part of our observational evidence.

To be sure, Russell's theory of descriptions has been criticized on the ground that it is not faithful to ordinary language. Strawson wants to insist that if one says 'Fido barks' the question whether there is one and only one thing which is Fido "doesn't arise"; in making the statement one merely takes it for granted that this is so.[6] Doubtless Strawson is correct in supposing himself to be closer to the spirit of ordinary language than is Russell. But in exact discourse, especially where the logic of nondemonstrative inference is in question, it will not do to employ singular terms in Strawson's manner. For, according to his view, in using the singular term one is taking for granted the truth of the empirical hypothesis that the singular term names exactly one thing. Yet this is a hypothesis involving a universal clause, and in a careful nondemonstrative argument

---

[5] Bertrand Russell, *Introduction to Mathematical Philosophy* (London, 1919), ch. xvi.

[6] P. F. Strawson, *Introduction to Logical Theory* (London, 1952), ch. vi.

one ought to bring this hypothesis out into the open (as the theory of descriptions does) so that its degree of confirmation can be assessed, lest the whole argument be vitiated. A singular term used in Strawson's manner ought not to occur anywhere in an exactly stated argument, and certainly not in the evidence.

If it be granted that our observational evidence ought not to be stated either in universal or in singular statements, then it seems to follow that it should have the form of existential statements. Existential statements, which assert that there are some things having such-and-such characteristics, sometimes do admit of being verified by observation. These statements are free from the difficulties to which universal and singular statements seem subject, so let us adopt them as having the form in which we shall think of our evidence as being expressed.

A further point about evidence deserves brief mention. Some people who take time very seriously seem to suppose it particularly important that evidence which confirms a hypothesis should pertain to events occurring after the time at which the hypothesis first was formulated. If predictions based upon a hypothesis come true, they think, this confirms the hypothesis far more than if the hypothesis merely accords with previously ascertained facts. But surely this must be dismissed as extravagance, for it is hardly plausible to suppose that the adventitious psychological fact of the hypothesis having been thought of prior to the obtaining of the evidence should make the evidence give stronger support to the hypothesis than it would have done had the items been thought of in reverse order. The support which evidence gives to a hypothesis is a logical matter and must depend upon the logical forms of the statements concerned, not upon accidental psychological facts. The question whether statement p confirms statement q is just as abstract as the question whether p implies q; the former is no more influenced by what anyone may happen to think or when anyone may happen to think it than is the latter. Whether or not these logical connections subsist between the statements involved is a matter which no temporal accident can influence. Predictions, then, deserve no special status in the logic of confirmation, and evidence statements do not all have to be dated. The general question is simply, under what logical conditions is one hypothesis better confirmed than is another with respect to a given body of evidence?

Three

Eliminative Induction

I

EVEN if we are able to agree about the character which the empirical hypotheses and the empirical evidence ought to possess, there remains the larger question of the principle which is to connect them. Suppose that we are given a body of evidence and a supply of hypotheses: what principle (or principles) ought we to appeal to when we seek to determine the respective degrees to which these hypotheses are confirmed or disconfirmed by this evidence? Fortunately those philosophers (as contrasted with gypsies) who have investigated the matter have practically all advocated principles of one or another of three main kinds —which we shall call eliminative induction, enumerative induction, and the method of hypothesis. It seems reasonable to suppose that some of these philosophers may be somewhere near right. So let us examine in turn these three main kinds of principles, trying to bring out the characteristic virtues and defects of each; for this ought to help us form a clearer view of what we ourselves should care to accept.

The majority of philosophers who have written about the subject have advocated one or another version of induction; they have thought of induction as providing the criterion determining the respective degrees to which any given body of evidence confirms various hypotheses✻An argument is called inductive, it will be remembered, if, from the evidence that a specified predicate is true of certain members of a class, it proceeds to a conclusion which is a generalization concerning the composition of the whole class with respect to that predicate or to a conclusion which is a prediction about whether this predicate applies to some particular unexamined member of the class. For example, the following arguments (whether or not they are good arguments) are inductive in form:

O'Reilly, O'Casey, Kelly, and Donahue are irascible Irishmen; so probably all Irishmen are irascible.

Among observed Texans, four out of five are observed to be millionaires; therefore, probably between 79 per cent and 81 per cent of all Texans are millionaires.

Lead, iron, and silver are metals and become superconductive at very low temperatures; so probably aluminum, being a metal, will be superconductive at such temperatures also.

Many writers have supposed that all valid nondemonstrative inference proceeds inductively and that all empirical knowledge both in science and in everyday life must be regarded as built up by induction from observational evidence. But before we can begin to ask whether this claim can be made good, we must inquire more exactly how the principle of induction is to be formulated.

✻There are two rather different ways in which induction might work. On the one hand, it may be said that induction proceeds

49

simply through the enumeration of instances—this would be induction by simple enumeration. According to such a formulation of the inductive principle, one generalization would be better supported than another just in case more instances in favor of the former (and of course none that contradicts it) had been observed. For example, if we have observed fifteen Irishmen to be redheaded (and have not observed any not to be) and if we have observed seventeen Irishmen to be irascible (and not observed any not to be so), then this inductive principle would enjoin us to regard the generalization that all Irishmen are irascible as more probable than the generalization that all are redheaded. Argument by simple enumeration, according to its proponents, is the fundamental mode of nondemonstrative inference; if we wish to justify belief in any empirical statement not verified by direct observation, then we must employ induction by simple enumeration. According to this view, there is no other way of constructing a cogent nondemonstrative argument, no other way of confirming empirical hypotheses.

On the other hand, it may be held that induction proceeds solely by elimination of rival generalizations. Some philosophers argue that the mere accumulation of instances cannot add any support to a generalization; only if there is reason for believing that these additional instances are different from one another in certain respects can they serve to increase the degree of rational credibility that attaches to the generalization. Indeed, common sense does suggest to us that, in order to establish the generalization that all swans are white, it does not suffice merely to observe that a large number of otherwise very similar swans are white; one ought rather to observe swans at different times of year, in different geographical regions, of different sexes, of different ages, and so on. One ought to observe swans which differ in as many respects as possible, for in this way one can hope to minimize the likelihood that it is some other characteristic possessed by the observed swans which (rather than the mere fact that they are swans) is the sufficient condition of whiteness. According to the proponent of induction by elimination, one should seek a variety of instances, and the differences among the instances are important because they serve to eliminate rival generalizations, only through the elimination of which can the generalization in question be established. Induction is viewed as a struggle in which the less unfit survive: a given generalization becomes better confirmed just insofar as its rivals are destroyed by being contradicted by the evidence. The proponent of eliminative induction would hold that, if all the evidence available were that fifteen Irishmen are irascible and seventeen redheaded, we could not legitimately conclude that either generalization is any more probable than the other. Lacking information about differences among the observed instances, this evidence, if it were all the available evidence, would provide very little support for either generalization, since it eliminates scarcely any rival generalizations at all.

It is striking to note how the climate of opinion has molded views of induction. Johnson,[1] Broad,[2] and Keynes[3] assumed almost as a matter of course that induction must proceed by elimination; they give only perfunctory reasons for this view. But more recent writers such as Reichenbach,[4] Carnap,[5] and Braithwaite[6] assume just as blandly that induction must work by enumeration; and they do not even argue against the eliminative view, for they scarcely seem aware that it might be

---

[1] W. E. Johnson, *Logic* (Cambridge, 1921–1924), pt. III, p. 24.

[2] C. D. Broad, "On the Relation between Induction and Probability," *Mind*, XXVII (1918), 389–404, and XXIX (1920), 11–45.

[3] J. M. Keynes, *A Treatise on Probability* (London, 1921).

[4] Hans Reichenbach, *Theory of Probability* (Berkeley, 1949), ch. xi.

[5] Rudolf Carnap, *Logical Foundations of Probability* (Chicago, 1950). See also his *The Continuum of Inductive Methods* (Chicago, 1952).

[6] R. B. Braithwaite, *Scientific Explanation* (Cambridge, 1953), ch. viii.

maintained. To be sure, in practice the contrast between these two opposed conceptions of induction is not so sharp, for usually any addition to the number of instances can be expected to increase the number of respects in which these instances differ among themselves. But the latent opposition remains and is significant from a theoretical point of view; if we are to understand the logic of nondemonstrative inference, we need to examine each of these views of induction more closely, in order to determine whether either of them can be regarded as satisfactory. Let us consider first induction by elimination, the method formerly the more popular.

II

It is all very well to claim that evidence ought to be sought which will eliminate rival generalizations; but how will this serve to make a given generalization any the more probable? Is it possible to explain eliminative induction in such a way as to show that it is reasonable to trust its results?

One often finds it being said, especially by persons of metaphysical bent, that induction can be trustworthy only if we are entitled to adopt some assumption about causality. This assumption usually is stated rather vaguely: 'Every event must have a cause,' or 'The future must be like the past,' or 'There is uniformity in nature.' Such formulations, however, are not very helpful unless they actually can be brought to bear upon inductive arguments. If, from the evidence that many different swans have been observed to be white, we are trying to argue to the inductive conclusion that all swans are white, it does not aid us to be offered the supplementary premise that every event has some cause or other or that there is some uniformity or other in nature. Such vague premises add nothing to our argument and are merely tiresome excess baggage. However, perhaps the idea which is moving obscurely in the backs of the minds of those who proffer such assumptions is that induction must work by elimination and that we need an assumption which will serve to guarantee that there is a limit to the variety of independent combinations of characteristics of things in the world. Properly stated, a principle guaranteeing some limitation upon independent variety might indeed be relevant to induction, for it might work to ensure that the eliminating of rival generalizations really could serve to increase the credibility of a given generalization. But just what should be expected of such an assumption?

Since we wish to know the conditions under which eliminative induction could make a generalization more probable, the mathematical theory of probability may be helpful here. Suppose we let 'g' stand for a generalization which we seek to establish (for instance, 'All swans are white'). Let 'e' stand for the inductive evidence which we have verified by observation (for instance, 'Some male swans are white; some female swans are white; some young swans are white; some old swans are white,' etc.). And let 'a' stand for the antecedent evidence we possess concerning the general nature of things. Now, if we use the solidus, as Keynes did, to express the probability function, we can write '$g/a.e$' to symbolize the probability of our generalization on the basis of our inductive evidence in conjunction with the antecedent evidence. Moreover, this probability is related to others, and according to elementary rules of the theory of probability we may assert that:

$$g/a.e = \frac{g/a \times e/a.g}{e/a}.$$

In words, the probability which the given generalization has on the basis of the antecedent plus the inductive evidence is equal to its probability on the antecedent evidence alone multiplied by the probability which the inductive evidence would have assuming the antecedent evidence and the generalization to be true, all divided by the probability of the inductive evidence on the basis of the antecedent evidence alone. Here, of course, we have assumed that each of the probabilities involved has a definite numerical value.

With this equation in hand, we can now consider more exactly what conditions must be fulfilled if eliminative induction is to lend support to the generalization. Ordinarily we may expect that our inductive evidence will be logically implied by the generalization in conjunction with the antecedent evidence: for instance, if the generalization asserts that all swans are white, then the antecedent evidence doubtless will affirm (among other things) that there do exist male and female and young and old swans, etc.; and this implies that some male swans are black, some female swans are black, etc. Thus if we assume a and g to be true, then e is certain, and its probability is one. That is:

$$e/a.g = 1;$$

whence,

$$g/a.e = \frac{g/a}{e/a}.$$

This last equation makes clear two conditions that must be fulfilled if eliminative induction is to work. In the first place, the generalization must have a finite antecedent probability; otherwise, no matter how much inductive evidence were gathered, its probability would always remain zero. In the second place, if the inductive evidence is to increase the probability of the generalization, then the antecedent probability that all this inductive evidence should be true must be finite also; moreover, if the gathering of more and more inductive evidence is to increase the probability of the generalization towards one (representing certainty) as a limit, then it must be possible for the quantity of inductive evidence to increase in such a way that the antecedent probability of its all being true approaches zero (representing impossibility) as its limit.

From the point of view of the theory of probability, the two conditions just mentioned must be satisfied if inductive evidence is to confirm the generalization. The mathematical theory of probability, however, does not give us any guarantee that these conditions ever will be satisfied. What sort of further assumption or principle could be introduced which would serve to guarantee that induction sometimes will work?

III

In answer to this question Keynes's suggestion is that we should adopt a postulate concerning the nature of the world, a postulate which he calls the "principle of limited variety." [7] This postulate, if we could adopt it as part of our antecedent evidence, would ensure that induction could proceed. Keynes's point of view is as follows: he suggests that we suppose all the innumerable properties (or at any rate all those properties concerning which we are going to make inductive generalizations) of the innumerable individual things in the world to be bound together in a limited number of groups, of a special sort. Actually he suggests that we regard all the other properties of things as caused by some finite set of "generator properties"; but the distinction between "generator" and other properties is inessential.[8] What matters is that every property should belong to at least one group and that there should be at most some definite finite number of distinct groups. This limitation upon the number of groups is a limitation of the independent variety in the world: for each group is to be such as to contain a distinctive finite subset of properties, and possession by any individual of

---

[7] Keynes, *op. cit.*, ch. xxii.

[8] See Jean Nicod, *Foundations of Geometry and Induction* (London, 1930), pp. 277 ff.

all the properties belonging to this finite subset is a sufficient (and necessary) condition for that individual to possess all the properties belonging to the group. Thus the occurrence of all properties of this subset will be an infallible symptom of the presence of the other properties belonging to the group.

Now, suppose that we are interested in some particular property Q and we wonder whether it belongs to a specific group of properties G. Since independent variety is limited, Q has to belong to at least one out of a finite number of groups, though in the absence of special information we have no reason for believing Q to belong to any particular group rather than to any other. In this case (according to Keynes) we may employ the "principle of indifference," which stipulates, roughly speaking, that probabilities are equal whenever there is no suitable reason for supposing them unequal. Thus we may say that Q is equally likely to belong to any one of the groups, and if there are at most n groups of properties, then the probability that Q belongs to G is at least equal to one divided by n. Now let us think of the group G as being a group to which the property P also belongs: there is a finite probability that Q belongs to G, similarly there must be a finite probability that P belongs to G, and therefore there must be a finite (though smaller) probability that both P and Q belong to G. Thus the principle of limited variety guarantees that there will be a finite probability that Q belongs to any particular group to which P belongs; and since P cannot belong to more than a finite number of groups, it follows that there will be a finite probability that Q belongs to every group to which P belongs. That is, there is a finite antecedent probability in favor of the generalization that all P's are Q's.

Thus the first condition that needed to be fulfilled seems to be taken care of here: the principle of limited variety seems to generate a finite antecedent probability in favor of any generalization which we seek to confirm. And furthermore, the observation of more and more individual instances (each of which we assume to be observed to differ in some respect from any instance previously observed) can progressively eliminate the possibilities of Q belonging to any group to which P does not also belong. In this way the accumulating of inductive evidence can raise the probability of a generalization so that it approaches certainty. Of course it is only eliminative induction which could work in this way; the principle of limited variety gives no support to the procedure of induction by simple enumeration, for the mere accumulation of instances of P's that are Q's will not contribute to raising the probability of our generalization at all. In order to contribute, every additional observed instance must serve to eliminate, with regard to one additional group, the possibility that P but not Q belongs to that group.

In order to make more plausible the basis for eliminative induction which this principle of limited variety provides, Keynes adds two further points. In the first place, someone might object that all the observed instances either exist at the present time or have existed in the past, and none of those that are going to exist in the future can yet be observed; and therefore it seems that the elimination cannot be anywhere near complete. Perhaps only past and present swans are white, while future ones are of a different color. Keynes attempts to disarm this objection by invoking a principle which he calls "the uniformity of nature." By this he means that the mere fact of a thing having one spatiotemporal location rather than another cannot be a necessary or sufficient condition for its having a property of any other sort. Keynes simply believes this to be self-evident. One might argue, however, that this principle is a truism reflecting the nature of the kind of language which we think it philosophically suitable to employ. For if singular terms are construed in accord with the theory of descriptions, one then can describe the spatiotemporal location of a thing only by specifying some of its properties or relations to other things; a mere spatio-temporal location simply as such would not be mentionable at all and so of course could not be said to be a necessary or sufficient condition of anything.

In the second place, Keynes points out that we are not obliged to assume the principle of limited variety to be true. It would suffice if we could assume that there is some finite antecedent probability in its favor. Then favorable inductive evidence (evidence indicating limited variety) could increase that initial probability. To have a finite antecedent probability in favor of the principle of limited variety would suffice to make it possible for us to use induction in order to raise the probabilities of generalizations, whether or not these probabilities could rise toward certainty as a limit; that might be satisfactory enough.

IV

In the preceding section we saw that Keynes's principle of limited variety could serve as a basis for eliminative induction. That is, if we were entitled to assign to this principle (on the basis of no inductive evidence) a degree of probability no less than some definite finite amount, then it would be possible for eliminative inductive evidence to confirm generalizations. The principles of the theory of probability (at least if we accept the principle of indifference) suffice to show that then any generalization concerning empirical properties would have a finite antecedent probability and that its probability could be increased by the addition of favorable evidence about observed individuals.

Now, no doubt one could criticize Keynes's formulation of the theory of probability; but quite apart from that, there are two serious difficulties about his view of induction. The first of them, which Keynes himself recognized clearly enough, is that no reason can be offered to explain why we are entitled to assign any definite antecedent probability to the hypothesis of the limitation of independent variety. What right do we have to consider this hypothesis to be credible? The hypothesis appears to be an empirical one; at any rate, it seems to assert something about the world, and it does not seem self-evidently true. Yet of course this hypothesis cannot derive its antecedent probability from any empirical evidence, if it is to serve its purpose. There seems to be no justification for believing it.

Suppose that we were to claim that the antecedent probability of this hypothesis about the limitation of independent variety is at least .1, or at least .01, or at least .001; but whatever figure we were to choose, our choice would be arbitrary and objectionable. And a similar difficulty arises with regard to the statement of the principle itself: the principle must assert that the number of independent groups of properties is not more than some definite finite number. But whatever number we choose, one thousand, one million, or one billion, our choice is arbitrary and objectionable. We need not even dwell upon the further difficulty involved in individuating properties. In any case, if we could formulate it, we should not know what minimum antecedent probability to assign to it, nor could we justify the opinion that the principle is credible to any specifiable degree. Keynes s philosophy of induction seems to require us to possess a priori information about matters of fact, information of a sort which we have no right to claim to possess. We reach an impasse from which there seems no escape.

Moreover, even if the various aspects of this first difficulty could be waived, the theory of eliminative induction still would be seriously defective on account of a second difficulty. This concerns the inability of eliminative induction to cope with many important arguments which occur in science and in everyday thinking. It will be noted that Keynes in giving his account of induction by elimination speaks only of properties, never of relations. His view apparently is that the language of science (and of empirical knowledge generally) is to be regarded as containing none but one-placed extralogical predicates, these to be observational predicates (a predicate is an observational predicate only if we sometimes can verify by direct observation that it is true of things), and that scientific knowledge will be statable in generalizations in which none but these extralogical predicates occur. But this, surely, is far too cramping a restriction. It is enough to point out that there actually are many important empirical hypotheses, both in science and in everyday thinking, which involve relations (that is, which need to be stated through use of predicates having two or more places). Consider the hypothesis that every object attracts every other; this we should probably want to regard as of the logical form '(x) (y) Rxy.' The hypothesis that every proton is heavier than any electron would most naturally be construed as of the form '(x) (y) $(Px.Ey. \supset Hxy)$.' And the hypothesis that every mutation results from a change in some gene would seem to be of the form '(x) $(\exists y)$ $(Mx \supset .Gy.Rxy)$.' Hypotheses such as these need two-place predicates; they are important hypotheses; yet they cannot be dealt with in a theory of induction such as Keynes's. Keynes cannot explain how such hypotheses have any antecedent probability, or how inductive evidence could increase their probability; nor does his philosophy of induction admit of any simple modification that would enable it to cope with hypotheses such as these. This is a serious defect in Keynes's philosophy of induction, for science would be crippled if it were obliged to abandon all such hypotheses as these.

Eliminative Induction

Surely no one can deny that Keynes gives a keen and illuminating account of eliminative induction. Yet it appears to have these unavoidable shortcomings. These objections seem to suggest strongly that induction by elimination cannot really be a satisfactory method upon which to base all nondemonstrative inference.

Four

Enumerative Induction

I

WE HAVE seen that there are two opposed conceptions of how induction ought to proceed, by elimination or by enumeration. The eliminative conception, examined in the preceding chapter, seemed to involve serious difficulties. The other conception, that of induction by simple enumeration, remains to be considered. We want to ask whether it is reasonable to regard induction by simple enumeration as a trustworthy mode of nondemonstrative argument, and moreover whether it is reasonable to regard it as the fundamental mode of nondemonstrative argument.

There are writers who have advocated induction by simple enumeration yet have taken their cue from Keynes when they try to explain its logic. That is, just as Keynes undertook to discover some assumptions about the world which, if we were entitled to believe them, would justify us in regarding eliminative induction as a reasonable procedure, so these writers seek to find some factual assumptions which, if we could know them to be true, would entitle us to regard enumerative induction as reasonable. Such assumptions, or presuppositions, of course would need to be credible even independently of whatever inductive evidence might be amassed in their favor—they would have to possess some antecedent probability. We might think of them (as Mill did [1]) as constituting a major premise attached to every inductive argument. Without this major premise, presumably, no inductive argument would be valid. But with this premise added, inductive arguments become valid in virtue of the laws of the theory of probability.

Russell, for example, has proposed a set of five "postulates of scientific inference," which he regards as fundamental to all nondemonstrative reasoning.[2] The first postulate, that of quasi-permanence, is stated as follows: "Given any event A, it happens very frequently that, at a neighboring time, there is at some neighboring place an event very similar to A." The remaining postulates, stated in similar language, assert that the world contains separable causal lines, that there is spatio-temporal continuity of causal lines, that similar structures ranged about a center usually have a common causal origin, and that analogies are usually trustworthy. Appeal to these postulates can suffice to justify both scientific and everyday inferences, Russell maintains.

A somewhat similar view has been adopted by Burks, who advocates a "presupposition theory of induction." [3] Any ordinary judgments about probability and confirmation ought to be construed as implicitly involving a set of factual presuppositions,

---

[1] J. S. Mill, A System of Logic, Ratiocinative and Inductive, 8th ed. (New York, 1874), bk. III, ch. iii, sec. 1.

[2] Bertrand Russell, Human Knowledge (New York, 1948), pt. vi.

[3] A. W. Burks, "On the Presuppositions of Induction," Review of Metaphysics, VIII (1955), 574–611.

according to this theory; and there is some one set of presuppositions, factual assumptions about the world, which serve as the foundation of our ordinary judgments about degrees of confirmation of hypotheses. Among his presuppositions Burks would include the temporal and spatial invariance of causal connections, the existence of causal connections, a principle of determinism, and other assumptions which he has not formulated explicitly.

This approach to the problem has a straightforward character, but of course objections can be raised. The difficulties to which this view of induction is subject are of the same kind as some that arose in connection with Keynes's views. For one thing, there appear to be many different sets of presuppositions that would justify induction; how are we to choose among them? For instance, in Russell's formulation, the vague words 'frequently,' 'usually,' and 'similar' could be made more precise, but if we made them more precise (by specifying how frequently, etc.) in various ways, then different sets of presuppositions would be obtained. It will hardly do to say that these various different sets of presuppositions are all equally good: for they are bound to lead to some conflicting estimates of probabilities. We should need to choose among these various possible sets of presuppositions, but we seem to have no grounds for making such a choice. For another thing, even if we could settle on one single set of presuppositions as most satisfactory, still there appears to be no good reason for claiming these presuppositions to be true. These presuppositions are factual statements about the world, and they do not look self-evident. It would be implausible for anyone to claim synthetic a priori knowledge of their truth. Thus it does not seem plausible to claim that our scientific knowledge must rest upon such a basis as this.

Enumerative Induction

II

To many writers, this problem about presuppositions appears impossible to solve. Therefore, a view like that of Reichenbach may come to seem attractive; for Reichenbach explicitly claimed to offer a theory of induction that would take nothing for granted.[4]

Reichenbach's theory of induction is based upon his frequency theory of probability. Rather than construing a statement such as 'Since it is cloudy today, there is a 50 per cent probability that it will rain tomorrow' as a statement exhibiting some logical relationship between the evidence 'it is cloudy today' and the hypothesis 'it will rain tomorrow,' Reichenbach construes it instead as a factual statement about the frequency with which rain actually does occur on the morrows of cloudy days. According to his view, in order to speak of the probability of rain tomorrow, we must envisage an endless series of morrows of cloudy days and we must think of ourselves as starting at the beginning of the series and observing longer and longer finite initial segments of this series and noting the relative frequency of rain in each of these finite segments. The relative frequencies of rain in these finite short-run bits of the endless series will themselves form an endless series. It may happen that, as we observe more and more morrows of cloudy days, the relative frequency of rain will settle around some definite value, say 50 per cent, which is approached as a limit. Then this limit of the relative frequency of rain in the infinitely long series of morrows of cloudy days is what is meant by the probability of rain on the morrow of a cloudy day. According to Reichenbach, a statement about probability is always to be construed as a

---

[4] Hans Reichenbach, *Theory of Probability* (Berkeley, 1949), sec. 91, and *Experience and Prediction* (Chicago, 1938), sec. 39.

statement about the limit of a relative frequency in an infinite series.

Reichenbach conceived the task of induction to be that of estimating the limit of a relative frequency in an infinite series such as this—all hypotheses in science are to be construed as probability statements of this sort. Now, we can examine only a finite initial segment of the series, yet we wish to estimate what the ultimate limit (if there is one) will be. If we have observed 1000 cloudy days and found that in just 500 cases there was rain on the morrow, we may then wish to adopt the inductive generalization that 50 per cent of all cloudy days are followed by rain, that is, that the limit of the relative frequency of rain in this infinite series is 50 per cent. Again, if we have observed 100 swans and all have been white, we may then wish to accept the inductive generalization that 100 per cent of all swans are white; meaning thereby that 100 per cent is the limit of the relative frequency of whiteness in an infinite series of swans. In effect the rule of induction which Reichenbach advocates is this: that if the relative frequency of the desired characteristic is such and such in the part of the series so far observed, then we ought to adopt this same percentage as our best estimate of the limit of this relative frequency in the infinite series; and the larger the number of observed cases, the greater weight is to be attached to this estimate.

This rule of induction seems natural, but what right have we to trust it, unless we may make some assumptions about the regularity of nature? Reichenbach's answer is that no such assumption is either justified or needed. Instead, he seeks to justify his rule of induction in the following way. This rule of induction, and this rule alone, can be relied on to lead us to the truth eventually, if there is any truth to be attained. For suppose we start out applying this rule on the basis of longer and longer initial segments of the series that we have succeeded in observing. At first, when we have observed only a small number of cases, our estimates of the limit of the relative frequency in the long run may be very wrong estimates, and they may continue to be wrong for a long time; but if there does exist a limit to the relative frequency, then sooner or later—by the very definition of a limit—we are bound to reach a point in the series after which our estimates will diverge less and less from the true value. If no limit exists, then no probability exists, and there is nothing to be discovered; and if the limit does exist, this rule of enumerative induction will enable us to estimate it to any desired degree of accuracy—provided we keep at the job long enough.

Thus, according to Reichenbach's justification of induction (the main idea of which has been embraced by a number of other philosophers [5]), this rule of induction is "our best bet." We can have no positive reason for believing that use of this rule will lead us to true conclusions, but we are entitled to assert that if there is any method at all that can do so this rule can. The contention is that this inductive rule occupies a very special, overarching status as contrasted with all other possible rules of nondemonstrative inference. Suppose, for instance, that scientists in their laboratories were not able to produce as reliable predictions as could be obtained by consulting old gypsy women: this might be so, Reichenbach would concede, but if it were so, then we should eventually observe enough instances so that the use of this rule of induction would lead us to estimate that in the long run gypsies have a higher relative frequency of success than have scientists; then we should incline

---

[5] For example, G. H. von Wright, *The Logical Problem of Induction* (Helsinki, 1941), pp. 178 ff.; R. B. Braithwaite, *Scientific Explanation* (Cambridge, 1953), ch. viii; J. O. Wisdom, *Foundations of Inference in Natural Science* (London, 1952), ch. xxiv; William Kneale, *Probability and Induction* (Oxford, 1949), sec. 44.

to trust the dicta of gypsies more than those of scientists. Thus the rule of induction gives us an overarching method which is "self-corrective" in that it leads us to trust whatever more specific methods actually show themselves to be the most reliable.

Reichenbach's view has the attraction that it seems to offer us a rule of induction which is fairly plausible in itself, and moreover which can be justified without our needing to make any factual assumptions at all concerning the nature of the world. Nevertheless, there are some difficulties.

In the first place, it is not too interesting to be told that induction is at least as good as any other method "if pursued long enough"; for "the long run" is a fiction without any definite empirical significance ("in the long run we are all dead" remarked the late Lord Keynes). Reichenbach's account of induction does not provide us with any guarantee that after any specified number of observations we are entitled to assume that our estimate of the long-run relative frequency will be within some specified degree of accuracy. Therefore, it would not seem that his abstract justification could serve to justify any concrete inductive argument. For instance, if I have observed a thousand swans and observed all of them to be white, I may wish to infer that probably all swans are white: and I want to know whether I am justified in making this inference. It does not help me much here to be told that if I continue using Reichenbach's rule of induction forever my estimates are bound eventually to approach the truth if the truth is to be had; this does not help because I cannot wait forever, and what I want to know is whether it is reasonable to accept this particular estimate here and now, made on the basis of the evidence actually available at present.

In the second place, Reichenbach has given the whole question of induction a curious twist from the beginning with his insistence that inductive generalizations must be regarded as statements about relative frequencies in infinite series. This has several untoward consequences. For one thing, it would oblige us to suppose that whenever one asserts an inductive generalization such as 'All swans are white' one is committing oneself to the untestable metaphysical assumption that there exist an infinite number of swans. Furthermore, according to Reichenbach's analysis, 'All swans are white' might be true—that is, the limit of the relative frequency of whiteness in the infinite series of swans might be 100 per cent—even though innumerable swans were not white (just as, if one considers the positive whole numbers in their normal order, the relative frequency of numbers that are not prime is 100 per cent in the long run, yet there are nonetheless infinitely many prime numbers in the series). Moreover, Reichenbach makes the meaning of an inductive generalization be relative to the arrangement of things in a series. Generally by rearranging the order of the items in a series one can alter the limit of a relative frequency in that series. Suppose, for instance, that we had a series of swans numbered 1, 2, 3, 4, etc. Suppose also that every even-numbered swan is white while every odd-numbered swan happens not to be white. Then if we consider these swans in their numerical order, we shall say that just half are white in the long run. But suppose we rearrange these same swans in a slightly different order: 1, 3, 2, 5, 7, 4, 9, 11, 6, etc. In this rearranged series, the relative frequency of white swans is one third. Thus, according to Reichenbach's view, it will make sense to assert an inductive generalization only if one specifies the order in which the items of the infinite series are to be arranged. And this seems unsatisfactory, for ordinarily we suppose ourselves to understand quite well enough what it means to assert that all swans are white without having decided upon any order in which an infinite series of swans are to be arranged.

Still other difficulties could be raised,[6] but these that have been mentioned suffice to make Reichenbach's philosophy of induction seem unsatisfactory. Is there any other more attractive way of defending enumerative induction without presuppositions?

III

These difficulties which beset Reichenbach's philosophy of induction are serious, but they still leave us room to hope that it might be possible to construct a theory of induction which would justify simple enumeration without needing to make any assumptions about the uniformity of nature. Such a theory, largely free from these difficulties to which Reichenbach's view is subject, has been proposed by Professor Williams.[7] His view is inviting in its simplicity, and it offers us what he takes to be a good reason for adopting and for trusting the method of induction by simple enumeration.

In his view, we must commence with the proportional, or statistical, syllogism. This is a mode of argument which, he contends, is basic to the theory of probability, being the one ultimate source from which numerical values for probabilities may be obtained (though of course once some numerical values have been introduced by means of the statistical syllogism, others may then be derived from these by means of the usual rules relating probabilities). The statistical syllogism is a mode of argument of the following form:

Of all the things that are M, $\dfrac{m}{n}$ are P.
a is an M.

Therefore (with a probability of $\dfrac{m}{n}$) a is P.

---

[6] See Max Black, *Problems of Analysis* (Ithaca, N.Y., 1954), chs. x and xii; Ernest Nagel, *Sovereign Reason* (Glencoe, Ill., 1954), ch. xiv.

[7] D. C. Williams, *The Ground of Induction* (Cambridge, Mass., 1947).

Enumerative Induction

Here m and n are integers, a is an individual thing, and 'M' and 'P' are empirical predicates.

As it stands, the statistical syllogism is not an inductive mode of argument; but it can be brought to bear upon induction in the following way. Let us suppose that we have observed n ravens and found m of them to be black, the rest not black. If n is a fairly large number, then using only algebra we can prove that, whatever the total number of ravens may be (so long as they are finite in number), the great majority of the n-membered subclasses of the class of ravens differ relatively little from the whole class in regard to the fraction of their members which are black. Thus we are provided with major and minor premises for this statistical syllogism:

> Of all the n-membered subclasses of the class of ravens, most differ little from the whole class in regard to the fraction of their members that are black.
>
> This sample, whose fraction of black members is $\frac{m}{n}$, is an n-membered subclass of the class of ravens.
>
> Therefore (with a good probability), this sample differs little from the whole class with regard to the fraction of its members that are black.

By means of straightforward algebraic considerations, the rough notions of "most," "little," and "good" could be replaced by exact algebraic formulations or by definite numerical values if m and n are specified. And no matter how large the class of ravens may be (so long as it is finite), "most" will mean a wholesomely large percentage, "little" a pleasingly small one, and "good" a fraction nearly equal to one, all provided n is fairly large. For instance, if n is 2500, "most" can mean at least 95 per cent, while "little" will then mean not more than 2 per cent, and "high" will mean .95. Furthermore, as n increases without bound, "most" approaches 100 per cent as its limit, "little" approaches zero as its limit, and "good" approaches one as its limit.

It is important to note the significance of the statistical syllogism just discussed. The major premise of that syllogism is not empirical; rather its truth can be certified by algebra alone. The minor premise constitutes enumerative inductive evidence: it sums up a set of observational statements gleaned from inspection of individual ravens. The conclusion of the syllogism implies an inductive generalization: 'The fraction of all ravens that are black differs little from $\frac{m}{n}$.' Thus the statistical syllogism enables us to pass from inductive evidence to an inductive generalization as our conclusion. The final conclusion is not so simple in form as are the conclusions of the form 'All P are Q' to which some other inductive methods might purport to lead us; but for practical purposes this sort of conclusion surely is satisfactory. Here no empirical assumptions about the world have had to be made; whatever the world may be like, this argument is impeccably and necessarily valid, provided we accept the statistical syllogism itself. We seem to be offered a strong reason for embracing enumerative induction as a trustworthy and fundamental method of nondemonstrative inference.

IV

Some critics have felt that the statistical syllogism is too good to be true and that it provides somehow too easy a way out of the difficulties surrounding the logic of induction. A few of these criticisms ought briefly to be noted, because of the light they may shed upon the issue at stake.

For instance, certain critics, taking very seriously the difference between past and future, have objected that this philosophy of induction is unsound because, though the world may have exhibited uniformities in the past, there remains a possibility that it may not continue to do so in the future, and thus inductive reasoning may go wrong. This objection is wide of the mark, of course, for the argument mentioned in the preceding section contains and need contain no factual presupposition whatever, and certainly none about the uniformity of nature through time or otherwise. It is just this freedom from such presuppositions which is its principal merit. To be sure, the conclusion of a statistical syllogism may be false even though its premises are true; but to point this out is merely to point out that the inference is a nondemonstrative, not a demonstrative one.

In the same vein, critics have objected also that this kind of argument fails on account of its refusing to employ any empirical evidence to show that the observed sample is a "fair" one. The sample could be biased, they argue, in which case this mode of argument would yield misleading conclusions. If all the black ravens were in the top of the urn, so to speak, it might be the case that nearly all ravens are not black even though all those observed are. Such an objection as this, however, likewise neglects the fact that the statistical syllogism does not purport to guarantee the correctness of its conclusion; if it did, it would be a demonstrative, not a nondemonstrative mode of argument. What the statistical syllogism does claim to show is that its conclusion is supported by its evidence. No more should be demanded; for to demand that the conclusion be necessitated by the evidence is in effect to claim that there cannot be any nondemonstrative arguments at all. It is enough, surely, that there should be no positive reason for supposing the sample to be misleading; given the evidence that we have, we must draw whatever nondemonstrative conclusions we can from it. To contend that given evidence cannot be employed in nondemonstrative argument unless there is further positive evidence that the given evidence is not misleading is to embark on a vicious infinite regress, a regress which would destroy the possibility of there being any valid nondemonstrative arguments at all.

There is a further point about the statistical syllogism which deserves notice, because it may seem to militate against the usefulness of this mode of argument. This involves the matter of infinity. We noted that the statistical syllogism is applicable only where the population concerned may be assumed to be finite in size. If the population were to consist of an actually infinite number of things, then the hyperpopulation consisting of all the subclasses having the same size as the given sample would likewise be an infinite class. In that case, it would be impossible to say that most of the possible samples closely resemble the population as regards composition; for while an infinite number of them might so resemble it, an infinite number of others would fail to do so, and no definite ratio could then be said to exist between the number of those that do and the number of those that do not. Thus the statistical syllogism cannot legitimately be employed except where we are entitled to assume the population to be finite.

It may be thought that this is a serious shortcoming. Indeed, some writers on induction have blithely asserted (along with Reichenbach) that populations of empirical things in the world may be, are, or must be actually infinite in number. Some like to fancy that the class of all swans that ever exist must embrace an infinite number of birds, that there are a limitless number of inhabitants of Africa, and so on. All who regard the spatio-temporal world as this full of things will have to regard the statistical syllogism as yielding an inadequate rule of induction, since it cannot serve to make probable any generalizations about such prodigious classes.

But is this prodigality justified? Need we seriously suppose that classes of things in the world may be so big? Surely this is a needless metaphysical assumption. For to say of any class of empirical things that it contains an actually infinite number of members is to make an assertion utterly devoid of empirical significance. This follows from the fact that no possible tests or observations could conceivably establish the statement; we might have empirical evidence that there exist at least a million swans, or at least a billion; but in the nature of the case, no evidence could establish that the number of swans is greater than every finite number. The supposition that the number of actual members of some class of empirical individuals is infinite is an untestable metaphysical supposition, and need not be taken seriously, so far as empirical knowledge is concerned. There are less things in heaven and earth than are dreamed of in some philosophies.

V

Provided the statistical syllogism be received, induction by simple enumeration does appear to be a method of argument whose soundness can be certified a priori. In this respect, it seems to enjoy an advantage over induction by elimination, which required that substantial empirical assumptions be made about the structure of the world. But is this apparent superiority a real one? Can it really be maintained that the statistical syllogism satisfactorily provides the desired demonstration of the soundness of induction by enumeration? And furthermore, can it be maintained that simple enumeration is the basic mode of nondemonstrative argument, through use of which all empirical knowledge may in principle be built up? Reflection reveals some difficulties here.

The statistical syllogism purports to provide a rule of enumerative induction enabling us to assess the degree to which the conclusion of any nondemonstrative argument is supported by its evidence. But a serious difficulty is that in some ordinary and important cases this rule is capable of leading to paradoxical results. Not infrequently we have to do with bodies of evidence which, if the statistical syllogism be uncritically employed, would lend high degrees of support to incompatible conclusions. Paradoxical results of this sort show that unguarded uses of the statistical syllogism and of the method of enumerative induction based upon it cannot be trustworthy. Consequently, doubt is cast upon the adequacy of this method as a foundation for all nondemonstrative inference.

The simplest sort of paradox of this kind arises when we have information about two classes to which an individual belongs. For instance, suppose that we have under observation a certain individual named Jones, whom we have observed to be a Texan and a philosopher; suppose, moreover, that from past inductions we know with practical certainty that 99 per cent of all Texans are millionaires and that only 1 per cent of all philosophers are millionaires. We can construct two statistical syllogisms:

> 99 per cent of Texans are millionaires.
> There is an individual named Jones who is a Texan.
> Therefore (with a probability of 99 per cent), this individual is a millionaire.

and:

> 1 per cent of philosophers are millionaires.
> There is an individual named Jones who is a philosopher.
> Therefore (with a probability of 1 per cent), this individual is a millionaire.

Here we start from a single body of evidence, yet we are able to construct two statistical syllogisms which assign incompatible degrees of probability to one and the same hypothesis. We have employed the statistical syllogism in an uncritical, straightforward way, and have arrived at a contradiction.

Enumerative Induction

Is there any way in which the statistical syllogism can be hedged round with restrictions which will prevent such untoward results? It will be seen at once that in our example the first syllogism takes as its premises only part of our total evidence, while the second syllogism takes another part, and this is the source of the difficulty. It seems natural, then, to invoke the "principle of total evidence" and to require that no nondemonstrative argument be employed which does not take into account all the relevant evidence. This plausible proposal would indeed prevent paradoxes of this sort from arising; it would prevent us from employing either of the above syllogisms, for neither of them takes account of the total evidence. But this proposal would do so at the cost of dethroning the statistical syllogism.

For suppose we insist upon the principle of total evidence and yet try to retain the statistical syllogism as our sole fundamental mode of nondemonstrative argument. We should then find that the statistical syllogism itself gives us no way of determining what among our evidence is relevant and what is not relevant to a particular conclusion: if we got involved in a paradox, then we should know that we had somehow neglected relevant evidence—but we should have no way of determining just which evidence really was relevant. In consequence, we should have to take the large number of statements that constitute our total body of evidence as the premises of each nondemonstrative argument that we sought to employ. But this would mean that our nondemonstrative arguments generally could not be in the form of statistical syllogisms: for by definition the statistical syllogism is an argument which must contain as premises only two statements (and these of specified logical forms); and generally our total evidence will consist of far more than just two such statements. It thus appears that the attempt to restore consistency by restricting the use of the statistical syllogism will succeed only if the statistical syllogism is restricted in a way which makes it impossible for it to be the sole fundamental mode of nondemonstrative argument. The rule of induction which the statistical syllogism yields at best would be applicable only in the unusual case where the total available evidence consists of just two statements suitable in form to be premises of a statistical syllogism—for only then could one claim to have an inductive argument of this sort not violating the principle of total evidence.

These reflections serve to suggest that the statistical syllogism provides us with a less satisfactory theory of induction than at first appeared. If we wish to defend induction by simple enumeration, we need to seek some further theory about its logic.

VI

The type of difficulty mentioned in the last section is not so artificial as the example given there perhaps suggested, nor is it restricted to the statistical syllogism. If we were to adopt any straightforward principle of induction by simple enumeration, we should find that difficulties of this type would plague us in a wide variety of familiar cases. We should continually be perplexed by opportunities for assigning incompatible degrees of probability to a hypothesis. This would happen whenever we sought to reason inductively about a case in which the inductive evidence points in two different directions.

Now, the sort of difficulty in question here is one which is obvious, and it has not escaped the notice of philosophers who advocate induction. They have seen that if they are to maintain that induction is the fundamental mode of nondemonstrative inference then they need to give some account of how inductive reasoning can be applied in cases where the inductive evidence is ambivalent. Reichenbach especially has sought to provide an explanation of this.[8]

---

[8] Reichenbach, *Theory of Probability*, pp. 430 f., and *Experience and Prediction*, sec. 41.

Enumerative Induction

Reichenbach reasoned as follows. If we think of our empirical knowledge as built up step by step, then we must think of the early steps as involving just a direct use of induction by simple enumeration. One observes a number of *P*'s a certain fraction of which are *Q*'s, and one then adopts the hypothesis that this same fraction of all *P*'s will be *Q*'s. However, later steps can proceed in a more complicated way. For once we have accumulated an initial supply of staightforward inductive hypotheses, we then can draw upon these hypotheses, modifying our later steps in the light of them. That is, we can make "second-order" inductions concerning the degree of successfulness of our "first-order" inductions. This procedure Reichenbach calls "concatenated induction." Reichenbach's claim would be that by appealing to concatenated inductions, one would be able to resolve those puzzles (at least in the cases important to science) about what to do when the inductive evidence points in contrary directions.

Concatenated induction would work in the following sort of way. Suppose that we were sociologists investigating the corruptibility of mankind. We might find (let us suppose) that hardly any clergymen or boy-scout masters (out of 1000 observed) can be influenced by bribes. But then we turn our attention to policemen. We observe, say, that in a number of cases Officer O'Malley, having stopped someone who deserves a traffic ticket, refrains from issuing it when offered a bribe of $5 or more. Officer O'Reilly in the cases observed refrains only if offered $20 or more. Officer O'Toole refrains only when offered $50 or more. Officer O'Hara, however, never is observed to accept a bribe, even though we observed 100 cases in which he is offered as much as $75. On the basis of this evidence we might frame the following first-order generalizations: in traffic cases, O'Malley can always be bribed by $5 or more; O'Reilly can always be bribed by $20 or more; O'Toole can always be bribed by $50 or more; but O'Hara is incorruptible. These first-order generalizations are reached by straightforward simple enumeration, based directly on the observed evidence. But what about the hypothesis that O'Hara is incorruptible? Ought we really to accept this hypothesis on the basis of the total evidence at hand? Not necessarily, Reichenbach would say; for we can make a second-order induction here. Consider the hypothesis that every policeman has his price: our first-order generalizations have provided three instances favorable to this generalization (O'Malley always has a price, O'Reilly does too, and so does O'Toole), and so we may argue that this second-order generalization is an acceptable one—and hence that O'Hara will be found always to have his price too, if the bribes go high enough. Thus appeal to the second-order generalization enables us to correct the conclusion which would have been reached by mere first-order induction. In this way, Reichenbach thinks, we work our hypotheses into more systematic form. We are freed from the obligation always to accept the results of first-order induction and are enabled to make our predictions in the light of previously accumulated generalizations.

Now, no one can doubt that in practice it often is necessary to take account not merely of the immediately relevant inductive evidence but also of indirect evidence which is relevant. But the question is whether a purely inductive theory of nondemonstrative inference can adequately explain the way in which this indirect evidence is relevant. A proponent of induction who seeks to explain this is bound to have to rely upon something very like Reichenbach's account of "concatenated inductions" (at least so long as he offers an informal theory of induction). But unfortunately Reichenbach's account is far from adequate. Reichenbach—in a way not untypical even of first-rate philosophers—seems to have been so assured of the correctness of his general doctrine (that all nondemonstrative argument is inductive) that he impatiently felt it needless to devote more than passing attention to the details of how it could be correct. Thus he failed to notice that his account of concatenated induction does not give us any definite criterion by appeal to which we can decide when to rely upon the second-level induction and when to rely upon the first-level induction. In the example of the last paragraph, the first-level inductive evidence (evidence that O'Hara has refused bribes) is favorable to the hypothesis that O'Hara is incorruptible; but the second-level evidence (that O'Malley, O'Reilly, and O'Toole are to be regarded as corruptible) is unfavorable. Clearly if we had only weak favorable first-level evidence and had strong unfavorable second-level evidence, then we should reject the hypothesis; while if we had strong favorable first-level evidence and only weak unfavorable second-level evidence we should accept it. But what criterion does Reichenbach give us for determining which evidence is the stronger? How are we to tell whether we ought to accept or reject the hypothesis? Reichenbach speaks of "weights" and "blind posits," but without giving any answer appropriate to a simple example like the one we have been discussing.

One might seek to answer on his behalf by saying that in a case like this we ought to reject the hypothesis if the unfavorable evidence involves more instances than the favorable evidence does and that we ought to accept the hypothesis if the unfavorable evidence involves fewer instances than does the favorable. Using this criterion, it might be said that the hypothesis of O'Hara's incorruptibility should be rejected since the evidence in its favor involves only 100 observations (O'Hara has been observed to refuse bribes on 100 occasions) while the second-level argument against it is based on hundreds of observations (observations of the corruption of O'Reilly, O'Malley, and O'Toole). However, the matter really is not so clear. For the second-level evidence might be taken as more comprehensive than this. Instead of making a second-level induction about

Induction and Hypothesis

the corruptibility of policemen, we might instead indulge in second-level reasoning about persons who serve the public as clergymen, boy-scout masters, or policemen; and more than 1000 observations have been made of such individuals, very few of whom have been observed to be corruptible. Thus our inductive prediction about O'Hara will depend upon what reference class we choose; and we have no clear criterion at all enabling us to decide what reference class is appropriate. The suggestion that we look to the number of instances pro and con does not seem to be a very helpful suggestion. It looks as though our reasoning is going to be arbitrary and chaotic so long as we rely upon a mere rule of induction; and this would seem to be the case for any theory of induction which is formulated informally.

VII

So far in our discussion of enumerative induction we have been considering rather informal theories; and we have observed that their somewhat loosely stated rules of induction tend to engender paradoxes and confusions. Perhaps, however, these difficulties could be avoided by a more rigorously formalized theory. This thought leads us to Carnap.[9] For Carnap's theory of induction does, in a certain sense, involve induction by simple enumeration; and it is the most elaborate and most fully formalized theory of the subject.

Working out a detailed version of the "Spielraum" theory at which Wittgenstein had darkly hinted, Carnap asks us to imagine that the evidence and hypotheses with which we have to deal all are stated in a definite formalized language. This language will contain a suitably formulated deductive logic, and its extralogical vocabulary will consist of a finite number

---

[9] Rudolf Carnap, *Logical Foundations of Probability* (Chicago, 1950).

of predicates plus a supply of singular terms each of which is to designate exactly one individual thing. The strongest possible kind of noncontradictory statement which can be expressed in this language will be one which affirms or denies each predicate of each individual in turn: such statements are called state-descriptions, and each state-description depicts a different "possible world." Any empirical statement that can be formulated in this language will be implied by some state-descriptions and will be contradicted by the remaining state-descriptions of the language. In order to obtain the results that are desired it is necessary not to regard all state-descriptions as having the same antecedent probability, for if they were the same the scheme would not work out satisfactorily. Instead, their probabilities are weighted in such a way that any two state-descriptions having the same structure are regarded as equally probable—two state-descriptions being said to have the same structure if and only if one can be obtained from the other simply by substituting some singular terms for others. Moreover, any two structures are regarded as equally probable: that is, the total probability assigned to the state-descriptions of one structure must equal the total probability assigned to the state-descriptions of any other structure. The a priori, or antecedent, probabilities of state-descriptions having been arranged in this way, we may then say that the a priori probability of any lesser statement is the sum of the a priori probalities of all those state-descriptions that imply it (metaphorically speaking, the sum of the probabilities of all the "possible worlds" in which this statement comes out true). The a priori probability of a statement is its probability without empirical evidence; if empirical evidence becomes available, the probability of the statement is likely to be altered. Carnap defines the probability of a hypothesis H relative to evidence E as the ratio of the a priori probability of the conjunction of H with E to the a priori probability of E alone; that is, the probability of H on E is the total probability of all those possible worlds in which both H and E come out true divided by the total probability of all those possible worlds in which E with or without H comes out true.

Carnap's scheme enables him to treat inductive arguments within an explicit logical framework; and he is able to show that in terms of his definitions certain kinds of inductive evidence can increase the probabilities of certain kinds of hypotheses. The difficulties of the sort discussed in the preceding section do not arise, for Carnap's scheme requires us always to consider the total evidence available; since we cannot leave out bits of evidence, and since his scheme shows us how in principle to manage total bodies of evidence, the puzzles arising from neglect of the principle of total evidence give no difficulty here. Instead of a casually formulated rule of induction which is ambiguous in its application, this more systematic scheme provides an unequivocal determination of the probability of any hypothesis with respect to any body of evidence in the language.

There are, however, several features of Carnap's theory which are less attractive. In the first place, Carnap follows Wittgenstein in assuming the availability of a language which contains one and only one proper name for every individual in the universe. From a philosophical point of view, this is a requirement which cannot be countenanced. Such a language at best could be available only to the Deity, and He presumably would have no need of empirical knowledge. We, unfortunately, do not know the exact number of individuals in the universe: nor could we possibly hope to form an opinion about how many there are—without resorting to inductive investigations.

Also, Carnap's theory requires that the primitive predicates employed in the language must all be logically independent of each other (if the statement 'All F are G' is necessarily true, then 'F' and 'G' are said not to be independent). Moreover, each place of a predicate must be independent of each other place (if 'Rxy' implies 'Ryx,' then the first and second places of 'R' are not independent, for instance).[10] These are tiresome restrictions, and they prevent the application of Carnap's theory to some languages that we should like to employ.

These difficulties are significant, but they are not irremediable. Kemeny has indicated in a striking way how these troublesome requirements may be circumvented.[11] He undertakes to do this by dropping Carnap's syntactical notion of a state-description and using instead the semantical notion of a model. What Kemeny means by a model for statements of a language is parallel to what some other writers have meant by an interpretation of statement schemata.[12] What is involved is an assignment of the individuals of an *n*-membered universe to the various extralogical primitives of the language. To each singular term of the language (if it contains any) is assigned exactly one individual; to each one-placed predicate is assigned some set of individuals chosen from the universe; to each two-placed predicate is assigned some set of pairs of individuals; and so on. To each one-placed predicate of next higher logical type is assigned a set of sets of these individuals; to each two-placed such predicate, a set of pairs of sets of individuals; and so on, until every predicate in the language has been provided with an extension. Since the number of predicates in a language (at least in any language with which we should care to deal) must be finite (and each of finite type), no universe containing only a finite number of individuals will yield more than a finite

---

[10] Y. Bar-Hillel, "A Note on State-Descriptions," *Philosophical Studies*, II (1951), 72–75.

[11] John Kemeny, "A Logical Measure Function," *Journal of Symbolic Logic*, XVIII (1953), 289–308.

[12] W. V. Quine, *Methods of Logic* (New York, 1950), sec. 24.

Induction and Hypothesis

number of different models for a given language. A model is said to be a model for a statement of the language if in it all these assignments are made in such a way that the statement comes out true.

Using this notion of model, Kemeny is able to define a measure function. Where M_n is the total number of models which a universe of n individuals will yield for the given language (or the total number of these consistent with the available empirical evidence) and where M_n^W is the total number of these models under which the statement W comes out true, we may write:

$$m(W) = \frac{M_n^W}{M_n}.$$

That is, the measure, with respect to universes of size n, of statement W in the language is the ratio of favorable models to the total number of models. Of course if we wish exactly to parallel Carnap's scheme, we may assign weights to the various models, equal weights to models having the same structure and the same total weight to every structure; then the measure of a statement will be the ratio of the total weight of the models in which the statement comes out true to the total weight of all the models.

It is now possible to go on to compare the measures of various statements. To say that $m(W_1)$ is greater than $m(W_2)$ will be to say that there is an integer j, such that for any integer k not less than j the value with respect to k-membered universes of $m(W_1)$ is greater than that of $m(W_2)$. Thus we are provided with a general technique for comparing the logical measures of statements in any given language. And obviously if we wish to do so we can employ logical measures in order to compare the degrees of confirmation of statements. For if one statement has a greater measure than another, then we could choose to regard it as the more probable—since the statement with the greater measure is the one which comes out true in more "possible worlds" (or in possible worlds of greater weight).

The difficulties mentioned, which afflicted Carnap's doctrine, are avoided by this approach of Kemeny's. There need be no restriction here to the effect that the primitive predicates of the language be independent. If some predicates are not independent, then all that happens is that some models will automatically be precluded which otherwise would have been available. Nor is there any need for the language to contain a proper name corresponding to each individual (as was necessary if we were to construct state-descriptions); indeed, the language does not have to contain any singular terms at all. And, as a further advantage, this technique deals in an obvious way with predicates of higher logical type, as Carnap's theory does not. Moreover, some of Carnap's somewhat counterintuitive results are avoided; specifically, a universal statement counts as more probable than a logically false statement. It appears that any desirable results and theorems obtainable in Carnap's theory could be obtained also using Kemeny's approach.

VIII

One troublesome feature of Carnap's scheme, even as modified by Kemeny, concerns universal generalizations. Generalizations will not admit of being confirmed by inductive evidence, according to this scheme. The reason is that we have in general no grounds for being entitled to assume that the size of the universe with which we have to do is less than some given finite number; we may suppose the universe to be finite, but we have no certainty about its maximum possible size. This being the case, a universal generalization will have to receive an a priori probability which is vanishingly small (on Carnap's view, zero; on Kemeny's view, greater than zero but less than any specific number greater than zero). And worse, no matter how many favorable instances be examined, the probability of the generalization still remains vanishingly small.

Carnap's way of meeting this is to abandon hope of confirming generalizations (such as 'All swans are white') and to concern himself instead with "instance confirmations," that is, with the probability of the prediction that the next swan will be white (or that any particular unobserved swan will be white). One is reminded of Mill's "reasoning from particulars to particulars." Carnap is able, on this principle, to establish a version of the Laplacean rule of succession; [13] thus, as the number of favorable instances observed increases without bound, the probability that a given individual, if it is a swan, is white increases towards one as a limit.

Carnap argues that we ought to rest content with this, that for practical purposes instance confirmations are all we need and that we ought to be satisfied to give up hope of confirming generalizations (and other universal hypotheses). No doubt half a loaf is better than none; and a good deal of empirical knowledge could perhaps be construed as consisting of nothing stronger than instance confirmations—for instance, an engineer wants to know whether this particular bridge will stand, and he does not much care whether all bridges of its type do so. But it would seem that in theoretical science universal hypotheses may have a more indispensable role to play; at any rate, it does seem somewhat paradoxical to maintain that no universal hypothesis ever can be confirmed. Surely it would be more desirable to obtain a scheme which could permit the confirming of generalizations and other interesting universal hypotheses, if such a scheme could at all plausibly be worked out.

Furthermore, if we adhere to the prejudice against singular terms which the second chapter urged, then instance confirma-

---

[13] Carnap, *op. cit.*, pp. 572 ff.

tions would not be possible at all: for Carnap's instance confirmation requires that singular terms appear both in the hypothesis being confirmed and in the evidence adduced in its support. It would appear that arguments involving instance confirmation must covertly employ assumptions about the existence and uniqueness of the individuals named, assumptions which according to this scheme cannot possibly be confirmed.

Another question arises in connection with hypotheses involving predicates of two or more places. Carnap's scheme does of course envisage hypotheses (and evidence) in which predicates having any number of places may occur. But the number of possible structures increases very rapidly when languages involving predicates with two or more places are employed, and the problem of how in general to calculate the number of structures compatible with any given statement becomes quite difficult.[14] Because of this complexity, Carnap has not yet dealt with such languages, but has considered only languages containing none but one-placed predicates. However, when one does come to consider more complex languages, it seems as though trouble may arise. For what can it mean to speak of the instance confirmation of a hypothesis which contains a two-placed predicate together with mixed universal and existential quantifications? For a generalization containing only one-placed predicates, the notion of instance confirmation makes clear sense: the generalization 'All dogs bark' has as an instance the statement 'Fido barks' (if we assume Fido to be a dog), and Carnap's scheme allows for confirmation of such instances as a practical substitute for confirmation of the generalization itself. But let us turn to a language containing some two-placed predicates and let us consider the hypothesis 'Every dog has a father,' whose logical form we assume to be '$(x)\ (Dx \supset (\exists y)Fyx)$.' We might at first suppose that statements such as 'Fido has a

---

[14] See R. L. Davis, "The Number of Structures of Finite Relations," *Proceedings of the American Mathematical Society*, IV (1953), 486–495.

father' would serve as practical substitutes for this hypothesis: but Carnap's scheme does not provide any way whatever of confirming the statement 'Fido has a father' except by direct observation of something that is Fido's father—no finite amount of evidence about other dogs could enable us to predict that Fido probably has a father, according to Carnap's scheme. Therefore, this sort of instance will not do. We might try instead to take as our instances statements such as 'Rover is father of Fido' (assuming both to be dogs); statements like this can be thought of as containing no quantifiers, and evidence about other dogs might, at least under certain very special circumstances, increase the probability of 'Rover is father of Fido.' However, the statement 'Rover is father of Fido' is so unrelated to the original hypothesis 'Every dog has a father' as hardly to be a practical substitute for it; moreover, it would seem that the kind of evidence which we should need in order to increase the probability of 'Rover is father of Fido' would be evidence that most observed dogs are observed to be fathers of one another—scarcely the sort of evidence which we should expect to obtain. It would seem that there is no kind of specific statement about a finite number of particular individuals which can serve as a usable substitute for a hypothesis like 'Every dog has a father.' Thus there may be an inherent difficulty in extending to richer languages the idea that our reasoning proceeds "from particulars to particulars." If this be so, then it may be impossible for Carnap's scheme to offer any sort of confirmation at all in connection with hypotheses containing two-placed predicates together with mixed quantifications. Since such hypotheses sometimes are important, this would be a definite shortcoming in Carnap's scheme.

Carnap's elaborate theory is an impressive and sophisticated one. Yet perhaps we are entitled still to hope for some other theory which might prove free from these troubles.

Five

Induction and Simplicity

I

IN THE last two chapters we have considered various theories of induction, eliminative and enumerative. Some of these theories have important virtues and advantages, yet each seemed to have its own characteristic shortcomings as well. The various difficulties which arise to plague these various theories prove for the most part nothing absolutely conclusive about whether or not an adequate theory of eliminative or of enumerative induction *could* be formulated. But they lend some weight to the presumption that it could not; for these various objections do not seem to be of sorts that could easily be overcome by minor modifications—they seem to suggest instead that we ought to seek some fundamentally different approach to the problem. This thought will be reinforced if we consider two further difficulties, general difficulties which involve the notion of simplicity and which apparently arise in connection with any theory of nondemonstrative inference taking induction to be fundamental.

The first of these two general difficulties arises in connection with generalizations about numerical quantities. In the examples of inductive arguments which we have so far considered, numbers have not figured very much. When numerical quantities are to be dealt with, we must employ a somewhat more complicated language, and the examples themselves tend to be more complex in their logical structure. For instance, Carnap in his theory of induction explicitly declines as yet to consider languages containing numerical locutions.[1] But arguments containing such locutions are important, especially so in science, where generalizations and predictions involving quantities continually need to be employed. Unless it at least gives hope of being adequate to deal with such quantitative hypotheses, no theory of induction can seem very plausible.

Any simply inductive theory, however, seems bound to lead to trouble when we attempt to apply it to quantitative examples. This can be illustrated by an example such as the following. Suppose that we have as evidence:

100 bodies falling freely for ½ second each fell 4 feet;
100 bodies falling freely for 1 second each fell 16 feet;
100 bodies falling freely for 2 seconds each fell 64 feet;
100 bodies falling freely for 4 seconds each fell 256 feet.

Reasoning inductively from this body of evidence, we might infer as probable the generalizations:

All bodies falling freely for ½ second fall 4 feet;
all bodies falling freely for 1 second fall 16 feet;
all bodies falling freely for 2 seconds fall 64 feet;
all bodies falling freely for 4 seconds fall 256 feet.

---

[1] Rudolf Carnap, *Logical Foundations of Probability* (Chicago, 1950), p. v.

Induction and Simplicity

Whence we might proceed to the further probable conclusion:

All freely falling bodies conform to the law, $s = 16t^2$. This line of inductive reasoning may seem to accord with common sense, yet it leads to difficulty.

The difficulty here is that, so far as merely inductive reasoning goes, we might just as well have inferred some quite different conclusion instead. We might have reached as conclusions:

All freely falling bodies conform to the law,

$$s = (16 + tan\ 2\pi t)t^2.$$

All freely falling bodies conform to the law,

$$s = 16\ (-1)^{4t}t^2.$$

We might have reached either of these as probable conclusions, or we might have reached any one of innumerable other conclusions which mathematical ingenuity could devise. In effect, our observational evidence provides us with a finite number of points on a graph, and our inductive generalization seeks to plot a curve through these points; but through a finite set of points innumerable different curves may be fitted. If we reason merely inductively we shall have no basis upon which to choose among these different curves, for each is consonant with the evidence, and there are the same number of observed instances in favor of each curve. Yet clearly we do need to choose among these alternative curves: for they yield quite different predictions regarding the as yet unobserved points on the graph, the as yet unobserved correlations of distance and time.

Some writers have spoken of inductive hypotheses as mere "conceptual shorthand" conveniently summing up masses of data. We can see how thoroughly misleading any such characterization is when we consider examples like the foregoing one. In this example there are alternative ways of generalizing from the data; but these are by no means equivalent, for they contradict one another with regard to as yet unobserved correlations of distance and time. These hypotheses obviously cannot be regarded as "conceptual shorthand." They are not just different ways of saying the same thing; they are competing hypotheses, incompatible descriptions, among which we must choose.

The difficulty is that a merely inductive philosophy of nondemonstrative inference gives us no basis upon which to choose among these competing hypotheses. We see that we must choose, that we cannot simply assign high degrees of probability to all these hypotheses, as a simple-minded adherence to mere induction might impel us to do. But no merely inductive rule of inference gives us any clue as to the criterion to which we should appeal when deciding how to choose. Reichenbach did once attempt to show that choice of the simplest curve can be justified by a purely inductive philosophy;[2] however, his argument is not very convincing, for he assumes without explanation that the derivatives of the curve we select ought to be continuous, and he fails to notice that innumerable curves besides the simplest will effect this.

In practice a scientist will unhesitatingly choose the simplest equation which fits his observations, other things being equal. He does not choose the simplest hypothesis merely because he finds it most convenient to do so—because the least expenditure of mental effort or of pencil and paper is involved. On the contrary, the scientist regards the simplest hypothesis as the one best supported by the evidence, most reasonable to believe, most probably true. This is evident from the fact that scientists repose confidence in predictions derived from the simplest hypothesis, but repose little or no confidence in predictions derived from unduly complicated competing hypotheses.

---

[2] Hans Reichenbach, *Experience and Prediction* (Chicago, 1938), pp. 377–380.

Induction and Simplicity

It would seem that we have one definite defect of any merely inductive theory of nondemonstrative inference. If we are to deal adequately with quantitative arguments of this sort, then we must import into our logic of nondemonstrative inference some principle about how to choose among competing hypotheses in cases like this. And presumably this must be some kind of principle of simplicity.

II

Another pervasive and important kind of nondemonstrative argument consists of those which have as conclusions hypotheses about the existence of unobserved things. Such arguments present a second general difficulty for induction.

Both in science and in everyday thinking, hypotheses concerning unobserved things have a large role to play. We believe that there is a cat behind the sofa, a house around the corner, works inside the clock; we believe that Hoboken, Antarctica, and the back of the moon now exist; we believe that certain events occurred in the past, that certain others will occur in the future. Our everyday beliefs are rich in hypotheses about the existence of things which we are not directly observing. Moreover, in science there are involved not merely hypotheses about particular things which are unobserved, but also hypotheses about whole classes of unobserved things. Hypotheses about atoms, electrons, valence bonds, genes, libidos, and so on come to play more and more vital roles in science as science becomes more advanced. An adequate theory of nondemonstrative inference surely must be able to deal with hypotheses about unobserved things. But can induction serve as a method for confirming such hypotheses? Let us not at this stage raise the question whether hypotheses about unobserved things ought to be eliminated from science (and construed as mere *façons de parler*). Supposing that we require such hypotheses, can induction handle them?

It is convenient to have a name for hypotheses of this character, so let us call them transcendent hypotheses. They are transcendent in the sense that they claim the existence of entities other than those which we directly observe, other than those which the evidence asserts to exist. From a formal point of view, these transcendent hypotheses can be subdivided into two kinds which differ in an important respect. One kind consists of hypotheses which can be stated so that all the predicates they contain are observational predicates; these hypotheses assert the existence of individual entities beyond those that are directly observed, but in order to do so they do not use any nonobservational predicates. Most of our familiar common-sense hypotheses about unobserved things can be construed as being of this kind, it would seem. And there can be no doubt that hypotheses of this kind do play an important role in our knowledge and that a philosophy of nondemonstrative inference must somehow account for them. The second kind of transcendent hypotheses consists of ones which can be stated only in such a way as to employ new predicates, nonobservational or theoretical predicates which never occur in evidence. Hypotheses of this kind hardly seem to occur in everyday thinking, but some philosophers would claim that in science hypotheses like these are essential. They would claim that such predicates as 'electron,' 'atom,' and 'gene' are not observational and that they cannot be defined by means of observational predicates, yet that science needs to employ hypotheses containing terms like these. If this should be so, then a satisfactory philosophy of induction would need to account for the confirming of these hypotheses too, it would seem.

We have two kinds of transcendent hypotheses that may need

to be confirmed. Let us consider how induction by elimination and by enumeration might fare in dealing with such hypotheses.

III

It would seem that any straightforward inductive argument which purported to support a hypothesis to the effect that there exists some unobserved thing would have to be an argument containing a relational premise. For instance:

Most Irishmen have children.
Kelly is Irish.
Therefore (probably) there are children who are Kelly's.

In this informal example it is the relation of parenthood which is involved. But whatever the example, some relation would be required in the major premise of the argument, and the argument would need to be of some such form as:

Most P's each bear R to some Q or other.
Some S's are P's (but none bears R to any Q as yet observed).
Therefore (probably) there are Q's (as yet unobserved) to which these S's bear R.

The proponent of induction may wish to contend that all cogent arguments about the existence of unobserved entities can be put into such form as this. But can the major premise of such an argument itself be established by inductive reasoning?

To begin with, we must note that a transcendent hypothesis of the second kind, one containing theoretical predicates, clearly cannot be confirmed by induction, whether that induction be enumerative or eliminative. For instance, if 'atom' is a theoretical predicate not definable by means of observational predicates, then we never can obtain observational evidence about atoms, and we never can establish any major premise of the desired sort: thus no inductive confirmation of a hypothesis about the existence of unobserved atoms is possible. Indeed, in general, an inductive argument may pass from evidence concerning some things known to be of a class to a conclusion about other members of the class; but it cannot pass from evidence about things known to be of one class to a conclusion about something known only to be of a quite different class. Whether we consider this a serious shortcoming of induction will depend upon how indispensable we think these theoretical hypotheses are to science.

Transcendent hypotheses of the first kind, ones containing only observational predicates, seem to be of more nearly unquestionable importance, however. Can eliminative induction or enumerative induction cope with hypotheses of this less abstract and more familiar sort?

The proponent of induction by elimination will be particularly hard put to explain how arguments of this sort can be carried through. For, as we saw in discussing eliminative induction, even the strong assumptions proposed by Keynes would suffice to make such induction possible only if we restrict ourselves to properties (admitting only one-placed primitive extralogical predicates to our vocabulary) and eschew relations (banning primitive extralogical predicates of two or more places). Whereas, in order to employ an argument such as that mentioned above, a major premise involving a dyadic relation must be employed (of course a relation of still higher degree could do instead). Such a premise, according to Keynes's scheme, would have no antecedent probability and could not be established by inductive evidence. Thus it seems that if one follows Keynes in regarding induction by elimination as the fundamental kind of nondemonstrative reasoning then one is debarred from giving any account of the status of hypotheses about unobserved things. It would seem that according to Keynes's view we never could have the slightest justification for supposing that anything exists which is not directly observed.[3]

Does enumerative induction fare any better? Can a theory of nondemonstrative inference which takes enumerative induction as fundamental explain how the first sort of transcendent hypotheses about unobserved entities can be inductively confirmed? Carnap's formalized theory is not helpful here, for the instance confirmations which it sanctions can only be hypotheses about objects observed to exist; so we must appeal to less formal theories such as those of Reichenbach or Williams. An argument by enumerative induction certainly can be constructed whose conclusion implies the existence of unobserved things; as was already indicated, its schematic form would be such as the following:

Most *P*'s each bear *R* to some *Q* or other.

Some *S*'s are *P*'s (and no *Q*'s to which they bear *R* are observed).

So (probably) there are some *Q*'s (not observed) to which these *S*'s bear *R*.

This kind of argument proceeds in an inductive way from its premises to the conclusion about unobserved things. But its first premise presents a difficulty. The premise that most *P*'s each bear *R* to some *Q* or other is itself an empirical hypothesis, requiring confirmation. Anyone who claims that enumerative induction is the sole mode of nondemonstrative argument must show how this premise itself can be confirmed by enumerative induction.

In order to provide inductive confirmation of the generalization that most *P*'s each bear *R* to some *Q* or other, what one would have to do would be to observe numerous *P*'s and to

---

[3] J. M. Keynes, *A Treatise on Probability* (London, 1921), pp. 299 f.

Induction and Hypothesis

observe Q's to which they bore R. Thus presumably the argument in full would have a form such as:

Numerous P's have been observed each to bear R to some Q. (1)
So probably most P's each bear R to some Q or other. (2)
Some further P's are not observed to bear R to any Q's. (3)
So probably there are unobserved Q's to which these P's bear R. (4)

Here premise (1), presumably based upon direct observation, gives inductive support to (2), an intermediate conclusion which together with the further observational evidence (3) supports the final conclusion (4).

But as it stands, this argument certainly is of doubtful cogency. Intermediate conclusion (2) is reached on the basis of the evidence of (1), yet (1) is not all the relevant evidence, for (3) is available also and surely is relevant. The information of (3) is unfavorably relevant to (2), for (3) informs us that there are P's which have not been observed to bear R to any Q's, and this militates against the hypothesis that most P's do bear R to Q's. In reaching the intermediate conclusion (2), this argument illegitimately neglects a relevant portion of the evidence.

It may be objected that (3) does not really militate against (2), because (3) does not contain evidence that any P fails to bear R to some Q—it merely says that some observed P's have not been observed to bear R to any Q's. But this will not do; for in the nature of the case, (3) constitutes the only sort of direct evidence unfavorable to (2) that we could expect. Observational evidence never could conclusively verify the hypothesis that a given thing which is P bears R to no Q whatever, since that would involve inspection of all Q's, knowing that none had been overlooked, and observing that this P

bears R to none of them. If (2) is an empirical hypothesis, it must admit of having unfavorable observational evidence brought against it, and (3) represents unfavorable evidence of the only sort there could be.

If we insist upon supposing that enumerative induction is our sole mode of nondemonstrative argument, then we are caught in an impasse here. (1) and (3) provide conflicting evidence. To be sure, it may be that the evidence provided by (1) really is somehow stronger; but the point is that no mere rule of induction allows us to adjudicate this conflict. A mere rule of induction of the kind advocated by Williams or Reichenbach gives us no way of evaluating the effect which the additional evidence (3) exerts upon the probability of (2). It would be legitimate for us to proceed with our argument only if we had evaluated this effect and found it to be sufficiently small. If we insist upon being inductivists, then we cannot evaluate the effect and we cannot legitimately proceed at all.

In everyday thinking we certainly do regard hypotheses about unobserved things as confirmable. If asked to justify this, we perhaps should appeal again to the notion of simplicity, claiming that sometimes certain hypotheses about unobserved things somehow are the simplest hypotheses that can be framed under the circumstances and that they therefore deserve to be regarded as credible. Here again the notion of simplicity suggests a way out of the impasse. But to admit that simplicity can be a factor of fundamental relevance to the confirmation of hypotheses would be to abandon the claim that induction is the sole fundamental mode of nondemonstrative reasoning. This would mean that the inductivist's philosophy about such reasoning would require fundamental revision.

IV

However, there remains a further line of thought of which the inductivist may wish to avail himself in attempting to show that induction can account for the confirmation of hypotheses about unobserved things. This line of thought depends upon the suggestion that a hypothesis may be confirmed by means of an induction concerning its consequents. The mode of argument is to rest on the principle that "if and only if all the consequents of a hypothesis are true, the hypothesis is true, while if all its tested consequents have been proved true, then probably all its consequents are true and so accordingly is the hypothesis." [4] Thus induction is to remain the basic principle, but, by making inductions concerning the consequents of hypotheses, we are to be able to confirm hypotheses of any required sort.

In considering this line of thought, let us note first that the only statements which are verifiable, strictly speaking, are observational statements. Now, if a hypothesis has a set of these observational statements as consequents, and if we obtain a sample of these consequents and by observation discover their truth-values, then it may seem that we can make an inductive inference as to the truth-value of the hypothesis itself. But there is obscurity here about what sort of consequents are in question. Are these to be consequents strictly deducible from the hypothesis, or may consequents be included that are merely probable in relation to the hypothesis? Only the former, surely, for the latter alternative would lead to a view which corrupts the understanding by considering the hypotheses as credible in proportion as they make the evidence probable rather than as the evidence makes them probable. Further, is an observational

---

[4] D. C. Williams, *The Ground of Induction* (Cambridge, Mass., 1947), p. 112. See also Keynes, *op. cit.*, p. 235n.

statement O to be regarded as a consequent of a hypothesis H only if H in isolation implies O? This would not be very satisfactory, since most ordinary hypotheses are such as not by themselves to imply any observational statement. Or is O a consequent of H provided there is some other already confirmed hypothesis which in conjunction with H implies O? This would merely defer the difficulty and besides would make the consequent merely probable in respect to the hypothesis. Or is O a consequent of H provided there is some other logically possible hypothesis such that O is implied by H in conjunction with it? Then the consequents would not even be probable with respect to the hypothesis, and any hypothesis would have as its consequents all possible observational statements.

In order to expound his scheme for inductive confirmation of hypotheses, the inductivist needs to claim that to each hypothesis there corresponds some distinctive set of observational statements. But this view is difficult to defend, as the above queries suggest.

Even if, as does not seem likely, these particular difficulties were resolved, others arise. For this method of employing induction at best could suffice to confirm only those hypotheses whose logically independent consequents are finite in number and are such that the conjunction of the consequents is equivalent to the hypothesis itself; and these are conditions which empirical hypotheses do not normally fulfill.

That the consequents would have to be finite in number is obvious if we are using the statistical syllogism as our mode of inductive argument; for the statistical syllogism would need to speak of the fraction of these consequents that are true, a way of speaking which makes satisfactory sense only with regard to a finite class. On the other hand, if we tried to employ Reichenbach's rule of induction or the like, we still should not be able to reach the desired result, even though Reichenbach's rule is geared to infinite series. For reasoning according to such a rule at best could lead us to estimate that in the long run 100 per cent is the limit of the relative frequency of true consequents among all consequents; yet this leaves open the possibility that innumerable false consequents exist even though the false ones be relatively scarce in comparison to the true ones. Obviously such a generalization is of no value to us; for in order to infer that the hypothesis itself is true, we need to be entitled to suppose that all—not merely 100 per cent in Reichenbach's sense—of the consequents are true. Thus we should have to limit ourselves to hypotheses whose consequents are finite in number.

Furthermore, not only must the verifiable consequents be finite in number, but their conjunction must be equivalent to the hypothesis itself: for if this were not the case, then it would be illegitimate to regard the truth even of all the consequents as any evidence in favor of the hypothesis. For example, suppose hypothesis H equivalent to the conjunction of a set of observational statements $O_1 \ldots O_n$, all of which are true. And suppose M is a false statement which is logically independent of H and of each of $O_1 \ldots O_n$ and which has no verifiable consequents. Now consider the hypothesis which is the conjunction of H and M; $O_1 \ldots O_n$ are the verifiable consequents of this conjunctive hypothesis and all these consequents are true; yet the hypothesis itself is false since M is false. This example shows that we must require a hypothesis to be equivalent to the conjunction of its verifiable consequents, if truth of the consequents is to constitute genuine inductive evidence in favor of the hypothesis.

Thus confirming a hypothesis by observing that some of its consequents are verified, and by then employing induction in order to argue in favor of the hypothesis itself, is a technique that can work only for a hypothesis which has no more than a finite number of verifiable consequents and which is equivalent to their conjunction. But to assert that hypotheses about unobserved things fulfill these requirements would be to adopt a radical kind of reductionism (a view to be discussed presently). In effect this would be to assert that hypotheses ostensibly about unobserved things really are not about unobserved things in any literal sense, but are translatable into statements about observed things only. This would contradict our premise that irreducible hypotheses about unobserved things are required by our empirical knowledge: the tentative premise upon which the arguments of this and the preceding section have been based.

The two general difficulties for induction discussed in this chapter make us wonder whether any inductive philosophy can be acceptable. At best, it would seem, induction could maintain its claim to be the basic mode of nondemonstrative inference only if two conditions be fulfilled. First, the inductive principle must be modified by the addition of some independent principle of simplicity so as to meet the difficulties with regard to numerically quantitative hypotheses. This doubtless can be done,[5] and the resulting theory might still be distinctively inductive in its character, although it might be an unattractively hybrid theory. Second, something must be done about hypotheses concerning unobserved things. Here it would seem much more doubtful that any mere addition could yield us a theory that would be workable yet still distinctively inductive in character. Here it would seem that if we wish to maintain the inductive point of view, we shall have to show that hypotheses about unobserved things are not literally necessary to our empirical knowledge but can in principle be dispensed with.

---

[5] See John Kemeny, "The Use of Simplicity in Induction," *Philosophical Review*, LXII (1953), 391–408.

Six

Reductionism

I

THE preceding chapter argued that one principal shortcoming of induction is its failure to yield any account of how hypotheses about unobserved things can be confirmed. If we accept this, then we are faced with a choice: either the claim that induction is the fundamental mode of nondemonstrative inference must definitely be abandoned and some different, noninductive mode of inference must be embraced that will enable us to confirm such hypotheses; or else it must be shown that we really can get along without actually needing to confirm any hypotheses which imply the existence of unobserved things. In this chapter and the next we shall investigate this latter alternative. First we shall consider the reductionist view, according to which transcendent hypotheses, although they are genuine statements, are not to be taken literally but are to be construed as shorthand for clumsier hypotheses about observed things only. Phenomenalism is a more specific version of this general view (phenomenalism being understood as the doctrine that only sense-data are directly observed and that all meaningful statements about unobserved entities are reducible to statements about sense-data alone).

In investigating the status of hypotheses about unobserved entities, let us mainly consider hypotheses about unobserved physical objects of familiar sorts—unobserved houses, tables, trees, cats; these seem to be unobserved entities of the most homely and natural kind. If it can be shown that we can make sense of our knowledge without supposing that hypotheses of this sort need be taken literally, then this will tend to establish the presumption that no hypotheses about unobserved entities of any kind need be confirmed; whereas, if we find that hypotheses about these familiar obejcts cannot be construed as mere *façons de parler*, then that will tend to show that such hypotheses are essential.

Traditionally in philosophy the name of realists has been assigned to those who believe that hypotheses about unobserved physical things must be taken literally. The realistic thesis (as was mentioned in the first chapter) is the seemingly common-sense one that physical things exist whether they are directly observed or not, that the whole spatiotemporal world of physical objects, and also whatever minds may happen to inhabit organisms in it, all together constitute a realm of real entities which exist in their own right and are not to be explained away. Some realists have held that no physical object ever is directly observed (that is, that just by looking, listening, etc., one may be able conclusively to verify some facts about one's sense-experiences, but that one cannot conclusively verify anything about a physical object), other realists have held that physical objects sometimes are directly observed; but they join in holding that among physical objects some at least do exist unobserved and that we often can obtain reasonably good evidence of their existence. Realists would say that ordinarily we take account of causal connections when we argue from our observations to the existence of unobserved things. For instance, if I observe that the needle of a compass is deflected when in proximity to a locked desk, this gives me some evidence on the basis of which to infer (with a moderate degree of probability) that there is a magnet in the desk drawer. I cannot directly observe the magnet, but because of its effects upon what I do observe, I may infer that it is there. And this in general, the realist may say, is the sort of way in which one can justify one's beliefs about unobserved things.

As we saw in the preceding chapter, arguments of this sort if valid cannot be merely inductive. A thoroughgoing empiricist may feel that the realistic thesis is thoroughly dissatisfying just on this account. Following Hume, he may ask, how can one possibly be justified in inferring that there exist real things beyond what one observes? One is able to verify statements only about what one does observe, never about things which one does not observe; hence, one never can verify that there is any constant conjunction between the existence of such-and-such unobserved things and the occurrence of such-and-such observations supposedly caused by them. It seems that the major premise must be unjustified in any inference which argues from the occurrence of certain observations to the conclusion that certain unobserved things exist. The realist's belief in unobserved physical objects at this stage appears to be based either upon mere animal faith or else upon some unjustifiable dogmatic principle about causation. Skepticism and dogmatism are equally unsatisfactory, yet realism seems to offer nothing else.

The question here, we must note, is not merely whether the realist's way of speaking happens to be more common or convenient or whether common-sense belief in the existence of the whole physical world is of some heuristic value to the practicing scientist. If we believe (as sensible persons cannot avoid be-lieving) that our empirical knowledge is more than a tissue of fictions, then we cannot be content to retain our belief in unobserved things and yet to regard it as an irrational "posit" having no justification and resting merely upon a psychological propensity. If there is to be empirical knowledge, it must be knowledge about real things, and it must be knowledge at-tained through valid reasoning. We must ask, are we or are we not entitled to believe that empirical reality extends beyond our observations of it? To what principles of nondemonstrative inference must we limit ourselves?

At this juncture what we may call reductionism presents it-self as an answer to these questions, an answer which purports to avoid the impasse to which realism seems to lead, an answer which purports to explain how empirical knowledge may be built up by means of induction without the need of any more elaborate principle. Since one can never be immediately ac-quainted with the unobserved things in the realist's world, one may wonder whether the existence or nonexistence of such things could ever be of the slightest significance. Perhaps all knowledge of empirical reality really concerns only observed things themselves, and nothing else beyond them. Along these lines, the proponent of reductionism may hope to undercut realism altogether by showing that the supposition that there are unobserved existents in the realist's sense is gratuitous as well as unjustifiable. The supposition also is senseless, he might then wish to add.

In order to undercut realism, the reductionist aims to show that instead of looking upon unobserved things as actual en-tities whose existence needs to be inferred we ought rather to regard them as logical constructs out of what we do observe. Now, a logical construct differs from an inferred entity precisely in that actual existence must be attributed to the latter but not to the former. An inferred entity is the real thing which one believes to exist in addition to those things the observation of which gives evidence of its existence; but a logical construct need not be regarded as having any real existence over and above that of the elements out of which it is constructed—to say something about a logical construct is just a covert way of saying something about the things of which it is constructed. For instance, common sense presumably supposes Julius Caesar to be an inferred entity, since from available historical evidence we conclude that there really was such an individual, even though we have not seen him. In contrast, the average American certainly is a logical construct, because in talking about him we do not suppose that there literally is any such person; we simply speak of him in order compendiously to convey generalizations about all existing Americans.

Reductionism is here being construed as consisting essentially in this attempt to interpret unobserved objects as constructs out of observed things. If the reductionist be correct, to make a statement about unobserved physical objects really is to make a statement that is about nothing other than observed things; consequently, one need not be skeptical about the possibility of confirming hypotheses about unobserved objects, nor need one accept any dogmatic inference beyond the realm of what is observed. For in observing what we do observe, we observe all that we have any right to suppose there is—all our talk about ostensibly unobserved things really being only shorthand talk about observed things. If it were to succeed, reductionism would have the advantage of allowing us to adopt a more parsimonious ontology and a simpler logic of nondemonstrative inference than would otherwise be possible. Our ontology could be more parsimonious in that we could dispense with belief in the existence of unobserved objects—the maxim that entities are not to be multiplied beyond necessity surely being a sound one for philosophy. And our logic of nondemonstrative inference could be simpler in that we should not need to adopt principles strong enough to justify the realist's style of inferences, some version of induction sufficing.

The motive behind reductionism thus is clear. It aims to undercut all the difficulties of realism by showing that the realistic inference from the observed to the unobserved ought never to be made. It is an attempt to strip away from the groundwork of empirical knowledge all superfluous, elaborate, dogmatic, nonempirical principles. It is an attempt to outline a minimum ontology, purged of needlessly metaphysical accretions, and with it a minimum logic of nondemonstrative inference. The purposes· of reductionism are exemplary. The serious question is, will it work? Does it make sense to think about the world in the reductionist fashion?

II

Before proceeding to a more detailed examination of the reductionist doctrine, it is worth while to mention a pair of objections that have been raised against it, objections which are instructive because of the ways in which they misconstrue the nature of the project which the reductionist is attempting.

The first of these objections, offered by a variety of philosophers, is to the effect that reductionism is unsound because it takes a wrong view of what one actually does observe. They object that the things we observe are observed to refer beyond themselves. Thus, if one looks at a book lying on a table, in a sense one does not directly observe that part of the table which is underneath the book; yet one can observe that the table as an enduring whole including that unobserved part nevertheless seems to be there. These objectors allege that reductionism tries to account for empirical knowledge of objects solely in terms of direct observations of the former sort and neglects these essential impressions of the second sort, these references to something beyond, which are bound up with our observations. If we simply pay attention to what we do observe, they tell us, we shall see that it contains implicit reference to real unobserved things.

Surely, however, it ought to be clear that an objection of this sort cannot hope to overthrow reductionism and establish realism in its stead. The sense in which the unseen part of the table appears to be there can be accounted for by the reductionist just as well as it can be accounted for by the realist; for the reductionist can explain this sort of impression as consisting in expectations of further observations of characteristic sorts; the unseen part of the table seems to be there in that one expects to observe it if one observes the book to be moved. The reductionist need not suppose that this impression involves any transcendent reference to objects that are not directly observed. This sort of objection fails, and it was bound to fail, for the objection is a phenomenological one, being based simply upon a description of the supposed content of experience as one finds it. But the issue between reductionism and realism is by no means one which can be settled just by appeal to a phenomenological inspection of the character of the evidence which experience provides; for reductionism as such need not be committed to any detailed view as to the character of what we directly observe. Rather, the issue concerns the relation of observations to unobserved things, an issue which can be settled only by an inquiry into the logical foundations of empirical knowledge. Of course we have the impression that we are observing a real world that contains unobserved things: but the question is, can the realistic thesis that unobserved things exist be justified? If there are questions that phenomenology can answer, this certainly is not one of them.

A second objection sometimes raised against reductionism

also misconstrues the issue at stake. This is the objection that reductionism illegitimately refuses to seek an explanation for the occurrence of what we do observe. Philosophers who object in this way say that we ought to ask what the causes are of the things we do directly observe, and we ought not to be satisfied with anything but a realistic answer. The reductionist, since he believes in no unobserved physical objects, ultimately will explain the occurrence of one observed thing only by correlating it with others, and the objection is that this is an inadequate sort of explanation. When it looks to me as if the compass needle is being deflected, I ought to ask for the physical cause of this phenomenon, the objector thinks; I ought to infer that there is a real unobserved magnet there in the drawer. After all, the odds are thousands to one against the observed phenomenon happening on the basis of mere chance, aren't they? The hypothesis that unobserved objects do exist and cause what we observe is the sort of hypothesis which alone really explains the occurrence of these observations, the objector contends, and therefore we ought to infer that such objects do exist. The reductionist's refusal to make this inference is just perverse.

But of course this objection merely begs the question. The realist has no right to assume blandly that the inference from observations to a realm of real unobserved objects is an inference which is legitimate. The thoroughgoing empiricist turns hopefully to reductionism just because he does believe that this procedure is wholly different in principle from the legitimate inferences of inductive science, a procedure far more dubious and unjustified than are inductive inferences. The reductionist would argue that a scientist observes many F's that are G's and thence infers that all or most F's are G's; but in the nature of the case, one cannot observe that observed things are accompanied or caused by unobserved things. Only if we make some elaborate and dogmatic assumptions (such as Russell's five postulates) would it seem that the inferring of the existence of the realist's unobserved objects could hope to be on a par with other inductive inferences. But assumptions of this sort are not truisms, nor can they be justified inductively; they would have to be synthetic a priori presuppositions, it would seem. The reductionist would prefer to do without such presuppositions; and if his reductionism should turn out to be a workable theory, it would dispense with them by construing scientific inference as having to do only with correlations among observed things, rendering gratuitous any attempt to argue in the realistic fashion. Thus this objection is a *petitio principii*, for it can carry no force unless we assume beforehand that reductionism is untenable; but that was just what the objection was supposed to prove.

III

Let us now proceed to a more concrete account of the reductionist's doctrine itself. Whatever we may have decided about the kind of things it is that are directly observed (whether these be sense-data or physical objects), we shall at any rate need to distinguish among three different kinds of statements. First, there are observational statements, statements of the kind which admit of being verified by direct observation. Second, there are statements of the kind which admit of being inductively confirmed on the basis of evidence which observational statements can provide. And third, there are transcendent hypotheses, which, even if they contain none but observational predicates, nevertheless do not admit of being inductively confirmed. The reductionist is satisfied with statements of the first and second kinds, but he dislikes statements of the third kind; he wants to avoid statements of the third kind and wants to regard all empirical statements as translatable into the first or second kind.

114

Reductionism

Among statements of the first two kinds, it is statements of the 'if . . . then . . .' form upon which those philosophers mainly rely who seek to explain how reductionism can work. Given a hypothesis ostensibly about an unobserved thing, for instance 'A cat is behind the sofa,' they would seek to analyze its meaning in terms of the various sorts of test that might take place which could yield evidence in favor of the hypothesis—if the sofa is moved, then what looks like a cat will be seen; if one reaches behind the sofa, then something warm and furry will be felt; if a dog is introduced, then strident feline sounds will be heard; and so on. All these various possible tests and their expected results would have to be stated in observational language, in terms only of the kind of thing that can be directly verified or inductively confirmed.

But according to what principle is this translation to proceed? The principle apparently is that every verifiable or confirmable 'if . . . then . . .' statement which could constitute favorable evidence for the transcendent hypothesis is to form an element in the translation of the latter. But how, then, are these various 'if . . . then . . .' statements to be joined together so as to constitute the translation? One proposal would be simply to join them by conjunction. But this would entail that the hypothesis about the unobserved object would be false if any single one of these 'if . . . then . . .' statements were false—this would oblige us to say that if one had reached behind the sofa without feeling anything warm and furry that would conclusively prove no cat was there, whatever the outcome of the other tests might be. But surely no single such test is conclusive; we do not want to believe that a transcendent statement would necessarily be false if just a single one of the 'if . . . then . . .' statements that could confirm it were false; scarcely any transcendent statement ever would be true in that case. A second proposal would be to say that the 'if . . . then . . .' statements should be joined just by disjunction. But this would not do either, because then the transcendent hypothesis would necessarily be true if at least one of the confirmatory 'if . . . then . . .' statements were true; and obviously this is unacceptable also.

A third proposal which is less implausible is that there be a series of conjunctions joined by disjunction.[1] According to this proposal, in asserting that there is a cat behind the sofa one would be asserting that all of a certain lot of 'if . . . then . . .' statements are true or all of another lot are true or all of another lot, and so forth. This structure of the translation presumably illustrates the fact that there are various different sets of tests, any one set being sufficient (if its results are wholly favorable) to establish the transcendent hypothesis. This seems the most promising proposal regarding the structure which the translation should have. The reductionist's thesis, then, is that any legitimate transcendent statement admits of being translated into a long statement of this structure compounded out of 'if . . . then . . .' statements each of which is the sort of statement that admits of being verified or inductively confirmed.

We have now an approximate account of the sort of translation advocated by the reductionist. But several vital obscurities remain, as the difficulties next to be considered will show.

IV

According to the pattern of translation suggested in the preceding section, the transcendent statement is to be regarded as equivalent to a disjunction of conjunctions of 'if . . . then . . .' test statements. The transcendent hypothesis is to be true provided all the components of at least one of these conjunctions of tests are true; the truth of all components of at least one of these conjunctions is a necessary and a sufficient

---

[1] A. J. Ayer, *Foundations of Empirical Knowledge* (London, 1940), ch. vi.

condition for the truth of the hypothesis itself. But how are we to decide what 'if . . . then . . .' statements ought to be included in any specific conjunction? Is there any criterion by which any verifiable or inductively confirmable statement of the 'if . . . then . . .' form can in principle be excluded from any such conjunction?

To return to our example, suppose that we wished to state just one conjunction of statements sufficient to establish the truth of the transcendent hypothesis, 'There is a cat behind the sofa.' As we saw, one might start by appealing to such tests as moving the sofa and seeing what was to be seen, reaching behind and noticing what was to be felt, or introducing a dog and listening for the result. However, this limited set of tests (even if all yielded favorable results) would not serve conclusively to eliminate the possibility that the actual situation might be quite different from that intended by the assertion that there is a cat behind the sofa. For instance, instead of a cat, there might be a small leprechaun behind the sofa, adept at imitating feline noises and disguised in well-fitting fur; if this were the case, the tests so far mentioned would yield favorable results yet the hypothesis would be false. Moreover, this small person would not be a nonempirical entity: other tests and observations not yet mentioned would enable us to check on his presence. These, however, might have to be far-reaching observations concerned with the general question whether creatures like leprechauns exist at all, and if so what their habits are. Such observations might have to be extensive and would carry us far afield. We should need to observe the curdling of milk in rural Ireland, aircraft failures in Thule, perhaps bewitchments in Basutoland.

But there seems to be no reason why a regress should not go on indefinitely here; consistent with any limited set of observations there could always be some physical state of affairs inconsistent with the truth of our hypothesis that the cat is there. And if we set about the task of systematically disproving every competing hypothesis, we should find that we had embarked upon a project which not only is endless but also proliferates wildly in all directions, all kinds of remote facts unexpectedly becoming relevant. Not only will we have to appeal to an indefinitely large number of tests, but, what is worse, it would seem impossible to mention any single possible test which is not in some way relevant. It would seem impossible to mention any single 'if . . . then . . .' statement (of the kind that admits of being verified or inductively confirmed) which we could be sure we were entitled to exclude from our conjunction. Ultimately we should find ourselves trying to include statements pertaining to all conceivable empirical events in the whole world throughout all time, any one of which must have some conceivable bearing on the matter. And this leads to two obvious objections against the reductionist's enterprise.

In the first place, the required translations cannot be carried through, because no finite conjunction of these 'if . . . then . . .' statements is equivalent to any transcendent hypothesis; and there is no such thing as an infinitely long conjunction.

In the second place, even if the first objection be waived, the translations still would not do, for every transcendent hypothesis would turn out to have just the same meaning as any other—a less than satisfactory result. This follows from the consideration that we cannot exclude any verifiable or inductively confirmable 'if . . . then . . .' statement whatever from any one of the long conjunctions; therefore, every transcendent hypothesis would have to be translated into one and the same unlimited statement. Should we confess, with the Absolute Idealists, that all assertions have the same meaning on the Highest Level of Reality? But this makes nonsense of our empirical knowledge and constitutes a second sufficient reason for rejecting such a scheme of translations. The whole question at issue was whether it is possible in principle for transcendent statements to be eliminated from our discourse. The reductionist contended that we could employ complex statements about observed things in their stead. But if, when we substitute the latter kind of statements for the former kind, our system of knowledge becomes an undifferentiated blur, then certainly the scheme of translation is not admissible.

V

There are some phenomenalist philosophers who would object to the conclusion suggested in the preceding section.[2] While admitting that indefinitely many observational tests are somehow relevant to any transcendent sentence, they would explain this relevance differently. They would claim that a transcendent hypothesis really can satisfactorily be translated into a finite compound of observational statements and that we need not become involved with these disastrously endless conjunctions. They would contend that a transcendent sentence is ambiguous and may be used to express any one of indefinitely many quite different statements, each one of which is equivalent to some finite compound of observational statements. When one utters the transcendent sentence, one really intends to assert only one among these indefinitely many statements, they would say. Because of its ambiguity, the transcendent sentence itself cannot be exactly translated into any compound of observational statements, but any specific statement that one might wish to make by means of the transcendent sentence can be translated in the desired way—so the reductionist's goal really is in sight.

For instance, the transcendent sentence 'A cat is behind the sofa' might be used to mean just 'If the sofa appears to be

---

[2] Roderick Firth, "Radical Empiricism and Perceptual Relativity (II),"
Philosophical Review, LIX (1950), 319–331.

moved, then it will look as though a cat is there, and if it looks as though a dog is introduced, then what sound like shrill feline noises will ensue.' Or it might be used to mean just 'If I seem to reach behind the sofa, it will seem that I feel a warm, furry object, and if I seem to stroke that object, it will seem to purr.' The same form of words, 'A cat is behind the sofa,' can be used to express these and other statements, each particular statement being translatable into some finite compound or other of 'if . . . then . . .' observational statements.

Now, in favor of this view we can admit that when one uses a transcendent sentence one is likely to have clearly in mind (if one has anything in mind) only a few explicit expectations about possibilities for further observations. But surely the reductionist goes too far who claims that these explicit expectations are all that is meant by the assertion, in the sense that it is necessarily the case that the statement asserted is true if and only just these expectations all are true.

In the first place, if transcendent sentences really functioned in this way, they would be doing a harmful rather than a useful job—by their ambiguity they would merely serve to veil logical distinctions from us. If this account of how transcendent sentences function were correct, one would be far wiser to dispense with such sentences altogether. One would then be left free to formulate one's empirical knowledge unambiguously in observational statements. Only a perverse penchant for extreme ambiguity ever could lead one to employ transcendent sentences at all, if this view of their function were correct. Yet surely transcendent sentences do have some more important role than this.

In the second place, this view about how transcendent sentences function does not seem to be compatible with communication and with scientific inference. The reductionist who adopts this view is contending that a transcendent sentence may legitimately be employed in order indifferently to express any one among innumerable different statements, these statements being (at least for the most part) logically independent of one another. But if everyday discourse is to make sense, there must be some firmness and objectivity in the meanings of the sentences employed; single sentences must express fairly definite statements, and the logical relationships among different sentences must be fairly fixed. But the reductionist is advocating ambiguity on a colossal and debilitating scale here. If sentences were used in this fashion, people would be unable to communicate with one another: for there could not usually be any way of guessing what a speaker meant if the sentence he uttered could equally well express any one among an infinity of logically independent statements. Moreover, the logical relationships which must obtain among scientific sentences if reasoning such as occurs in scientific treatises is ever to be correct would be dissolved by the radical ambiguity which this sort of reductionism advocates; it would be impossible to maintain that one scientific sentence does or does not imply another, since each sentence concerned would admit of being construed as expressing any one among infinitely many independent statements.

These considerations suggest that it would not be very satisfactory to accept the view that these reductionists propose. This version of reductionism is really no more tenable than that previously considered.

VI

The pattern suggested in the penultimate section for translating transcendent hypotheses proved unacceptable, and we have seen that the attempt to rectify it by requiring the translations to be finite is not satisfactory either. Objections parallel to those that have been raised would apply against any other arrangement of truth-functional connectives for joining together the 'if . . . then . . .' statements (and would apply even if statements of other forms than the 'if . . . then . . .' form were employed). Indeed, any truth-functional statement is reducible to the normal form of a disjunction of conjunctions, just the form we considered. Furthermore, it seems unwarranted to suppose that any connectives other than truth-functional ones could be used to join together the various statements so as to constitute a translation of the transcendent hypothesis. Such relations as implication, equivalence, entailment, or probability are intensional relations in the sense, roughly speaking, that whether they hold between any given statements must be certifiable a priori and depends not upon matters of empirical fact but only upon logic, in some broad sense. Such relations would be of no use in joining together the 'if . . . then . . .' statements so as to form a translation of a transcendent hypothesis, because the hypothesis is empirical and its truth or falsity must depend upon empirical, not merely upon a priori considerations.

We may note also that the unanalyzed relation which has been called "matter-of-fact connection" [3] (and which is said to be akin to the relation of causal necessity) could hardly serve as a relation between these 'if . . . then . . .' statements. Even if it makes sense to speak of "matter-of-fact connection" as a relation between events or attributes, surely it is not also a relation between statements about events or attributes. And in any case, to espouse such "real connections" would involve committing oneself to belief in an ontology of real possibilities: entities whose existence would be at least as difficult to confirm as is that of the realist's world of actual unobserved things.

The versions of reductionism previously considered did not seem satisfactory, and these considerations suggest fairly strongly

---

[3] C. I. Lewis, *An Analysis of Knowledge and Valuation* (La Salle, Ill., 1946), ch. viii.

that there is no likely way in which reductionism can be modified so as to become a tenable view.

VII

The preceding sections have attempted to present as favorable a case for reductionism as can justly be made out and then have tried to elicit some of its shortcomings. In concluding this account, let us notice two final difficulties that are involved.

The first of these difficulties arises in connection with the view to be taken of the minds of other persons. What does it mean for me to attribute thoughts, sensations, or feelings to another individual? Do I literally mean that this individual is enjoying mental states which though I cannot directly observe them are nonetheless similar in kind to the thoughts, sensations, and feelings which I myself enjoy? Or when I attribute these mental states to another, am I really just making some covert predictions about what may directly be observed concerning his behavior? If reductionism is to remain true to its thesis that transcendent hypotheses should be construed as translatable into statements about what is directly observed, then the reductionist must claim that the latter, not the former meaning is what is essentially involved in such assertions about other minds. The reductionist may perhaps wish to maintain that one can directly observe one's own feelings, but he will have to maintain that one must view the minds of others in the behavioristic fashion, because at best one cannot directly observe anything of another person but his behavior. Statements about the feelings of others are transcendent hypotheses and must be translatable into statements about what I can directly observe.

Some phenomenalistic reductionists indeed profess not to accept this view of the matter.[4] But insofar as they do not, they

---

[4] Ayer, *op. cit.*, ch. v.

are being untrue to the radical empiricism which alone can lend a veneer of plausibility to their doctrine. For the belief in other minds—the belief, that is, that statements about other minds are not mere *façons de parler*, mere ways of describing directly observable behavior—is an eminently realistic belief. It is a belief in the existence of entities whose existence I cannot verify by observation or confirm by mere induction. And if one is going to accept belief in other minds, belief of this realistic sort, why then should one be unwilling to believe in the existence of unobserved physical objects beyond the range of one's observation? Only a thoroughgoing reductionism has any right to claim our sympathy; unconvincing indeed are the half-measures of the quasi-reductionist who regards physical objects as having the status of mere logical fictions yet who admits other minds to be irreducibly real. In order to give an account of the justification of his beliefs about other minds, he would have to rely upon principles of inference at least as strong as those required in order to justify belief in the existence of unobserved physical objects, and if he is going to accept such principles, he might as well be a thorough realist.

The consistent reductionist, who is unwilling to be a realist at all, will have to adopt behaviorism as his view about the meaning of assertions concerning the mental states of others. He will have to hold that to make a statement about the mind of another is just to make some complex statement about what one directly can observe concerning his behavior. But this doctrine does not recommend itself. It is not the sort of doctrine that one could seriously adopt, for a person who took it seriously would be unable to make sense of his moral convictions. Everyone who is not a moral imbecile believes himself to have an obligation not to inflict suffering upon others wantonly. But if the mental states of others do not exist (except as *façons de parler*), then it cannot make sense to say that I am under an obligation not to cause suffering to others. If the suffering of others were a mere logical fiction, then morally it ought to be quite all right for me to go about torturing infants with red-hot irons—they might writhe and scream a bit under such treatment, but surely it does not make sense to say that just by causing writhing and screaming one is guilty of causing anything intrinsically bad. This is only a dialectical argument; still, the ineradicable sense of being obligated to consider the feelings of others—a sense which cannot be dismissed as irrational— must militate against reductionism.

In addition to this matter of other minds, there is a second general sort of difficulty to which reductionism is subject. This concerns the nature of confirmation. The reductionist has tried to analyze the whole significance of a transcendent hypothesis in terms of some special set of tests by means of which it would be confirmed. But he fails to notice that it never is merely a single hypothesis that is at stake in a particular empirical test but rather a whole body of hypotheses; if the test yields an unfavorable result, this merely shows that something is wrong with this body of hypotheses—it does not show which particular hypothesis is false. For example, how is one to decide what observational evidence would confirm the hypothesis that a cat is behind the sofa? Only if one accepts the empirical generalization that cats mew will one regard a mewing sound as confirmatory evidence. Only if one accepts the generalization that cats have tails will one regard as confirmatory the evidence that there is a tail protruding from behind the sofa; if one held some different hypothesis about the anatomy of cats, this evidence might be regarded as disconfirmatory instead.

These considerations suggest that the attempt to take a single isolated hypothesis about some unobserved thing and to translate it into test statements is an attempt that is misguided from the start. It never is just one such hypothesis alone that has any definite implications for experience; rather such a hypothesis acquires such implications for experience only when it is entertained against the background of a body of more or less accepted hypotheses about the physical world. Small wonder, then, that when the reductionist considers isolated transcendent hypotheses and tries to assign to each a distinct set of experiential implications he finds that in the end all turn out to mean everything, none having any distinctive significance of its own. Where no body of information about the world is presupposed (the reductionist not being entitled to presuppose any information about a world of unobserved entities), all reductionism becomes incoherent.

Thus we see that the relation between transcendent hypotheses and observed things must be looser than the reductionist allows. What one has to do is to relate a whole system of hypotheses about unobserved things to a whole set of observational statements. We cannot justly say that any one statement in the former set strictly corresponds to any specific statements in the latter set. Rather the occurrence of new and unexpected observations merely requires us to make some change or other in our system of beliefs about unobserved things, and we make whatever adjustment somehow is simplest and most coherent. There is no one-to-one correlation between elements in the two realms, but instead a give-and-take, a balancing of one whole system against another, a confirming of the whole body of transcendent hypotheses by the whole body of observational evidence. Our particular statements about unobserved things "are established on the presupposed basis of the whole normal world from which it is not logically nor factually correct to disassociate them," Bosanquet says; [5] or as Quine puts the point, "they

---

[5] Bernard Bosanquet, *Implication and Linear Inference* (London, 1920), p. 64.

face the tribunal of sense-experience not individually but as a corporate body." [6]

Initially reductionism may have seemed attractive, for by promising to eliminate transcendent hypotheses it offered the hope that we might be able to get along without needing any troublesome presuppositions or any more elaborate mode of nondemonstrative argument than induction alone. But if these various objections are cogent, then reductionism cannot make good its promise.

---

[6] W. V. Quine, "Two Dogmas of Empiricism," *Philosophical Review*, LX (1951), 38.

Seven

Formalism

I

WE HAVE now examined the reductionist's view that all empirical statements ostensibly about unobserved entities ought to be construed as really being statements about observed entities only. If this view could have been maintained, it would have meant that in principle all our empirical knowledge could be couched in discourse purged of any commitment to the existence of unobserved entities; and this in turn would mean that we could avoid the difficulties posed by the seeming need for special principles of nondemonstrative inference to confirm hypotheses about such entities. But the reductionist's view does not turn out to be tenable: for if we resolutely followed his recommendation and translated all our statements about unobserved entities into statements about observed entities only, then most of our everyday discourse would become incoherent and nonsensical. The resolute reductionist who really refuses to believe in the literal existence of unobserved objects ought to leave off making statements about them altogether, if he does not want to talk nonsense. The rest of us will not wish to stop making these familiar and indispensable statements about unobserved things, statements so essential both to everyday discourse and to science; instead we shall prefer to reject the reductionist's thesis.

This result naturally is dissatisfying to the thoroughgoing empiricist, who wants to defend science and at the same time wants to avoid commitment to anything that seems to smack of the nonempirical, as do unobserved entities and as do the principles of nondemonstrative argument that would seem to be required in order to justify hypotheses about them. Such an empiricist is not easily dissuaded; he may admit that the reductionist's thesis cannot be maintained, but he may still hold to the idea that these seemingly nonempirical elements must be eliminated from the groundwork of empirical knowledge. Now, one line of thought which some philosophers have suggested in this connection is that hypotheses about unobserved things (especially such things as atoms or molecules) be regarded merely as hypotheses about the "logical structure" of those unobserved things; [1] these philosophers have supposed that we cannot confirm hypotheses about the nature of unobserved things, but that it will not be too difficult to confirm hypotheses about their "logical structure." Unfortunately this view is not very helpful, for to say of an object or part of the world that it has such and such a "logical structure" is to say nothing at all, in view of the fact that this notion of structure is so abstract as to allow that any two sets of things whatever have the same

---

[1] For example, Bertrand Russell, *Introduction to Mathematical Philosophy* (London, 1919), p. 61; William Kneale, *Probability and Induction* (Oxford, 1949), p. 94.

logical structure in some respect or other.[2] However, there is an allied line of thought, which may appear more satisfactory. This is the thought that sentences which purport to mention unobserved entities perhaps ought not to be regarded as expressing anything true or false at all, that instead they ought to be regarded merely as inscriptions, strings of marks, which can be manipulated in useful ways. This sort of view has been called "marksism" by Professor H. M. Sheffer; a more common name for it is "formalism."

Formalism began as a view of the nature of mathematics and logic. Doubtful that it could be philosophically legitimate to regard the formulas dealt with by logic and mathematics as being statements capable of truth or falsity (for if they were statements, what evidence would we have in their favor; might not some of them have to be regarded as synthetic and a priori?), some logicians and mathematicians came to suppose that their systems really have to do only with strings of marks of various sorts. According to this view, it is the task of the logician to investigate ways in which marks of certain shapes can be derived from others, provided certain rules of the game are adhered to. A logical or mathematical system, like the magnified game of noughts-and-crosses, will consist of a certain repertory of marks which may be combined and manipulated according to certain rules so that various new configurations are obtained as results. When such a game with marks is played, the strings of marks which appear at various stages in the play are nothing more than inscriptions; they have not been assigned any meanings, so they are not statements, they do not assert anything, they are neither true nor false. To be sure, at certain stages in some of these games with marks, these "calculi" as

---

[2] See H. J. McLendon, "Uses of Similarity of Structure in Contemporary Philosophy," *Mind*, LXIV (1955), 90.

Braithwaite calls them,[3] there may occur inscriptions which do admit of being interpreted in useful ways; and if we wish, we may assign meanings to these particular inscriptions and may interpret them as expressing statements—this will make possible some practical applications of the otherwise purely uninterpreted system. But it is uninterpreted systems with which the pure logician or pure mathematician is concerned, and therefore he is freed of any need to worry about whether the formulas with which he busies himself are true. Those who hold the formalist view believe that by construing logic and mathematics in this way they succeed in avoiding puzzling difficulties about the status of logical and mathematical truth, difficulties which seemed to present pitfalls for empiricism.

This sort of view of logic and mathematics has had a good many proponents in recent years, and it is not surprising that some should seek to construe troublesome empirical hypotheses in this same formalist fashion. This is especially so with regard to hypotheses such as occur in advanced scientific theories. Many philosophers nowadays do hold that science must employ hypotheses about unobserved theoretical entities (that is, hypotheses containing predicates which do not admit of being defined by means of observational predicates). Hypotheses about atoms and electrons or about genes and libidos might be examples: many philosophers would claim that such hypotheses are indispensable to science yet that terms like 'atom' and 'gene' are theoretical predicates which cannot be defined by means of observational predicates. But obviously hypotheses of this character cannot be confirmed by straightforward inductive reasoning, since we never succeed in observing any theoretical entities and hence never possess inductive evidence upon which to base hypotheses about them. Philosophers who believe these

---

[3] R. B. Braithwaite, *Scientific Explanation* (Cambridge, 1953), p. 23.

hypotheses necessary to science are especially likely to embrace a formalist view, for they want to retain these hypotheses but without being obliged to confirm them. They are likely to wish to construe these hypotheses about unobserved things as mere formulas of an uninterpreted system, formulas that are not to be regarded as expressing conjectures that one could literally believe or disbelieve but that are merely useful strings of marks one may employ.

If the formalist point of view were to suffice to account for hypotheses containing theoretical predicates, then one might expect that by a natural extension it could be made to account also for the less abstract sort of hypotheses about unobserved things, that is, hypotheses which contain only observational predicates yet which imply the existence of entities distinct from any that have been observed. We might hope to be able to construe these hypotheses involving the existence of unobserved things of familiar sorts—hypotheses obviously indispensable both in science and in everyday life—as formulas not requiring inductive confirmation. And we might hope then to be able to maintain that no principle more elaborate than mere induction need be employed in nondemonstrative argument.

Of course, if the formalist view of empirical hypotheses is tenable at all, one might wish ruthlessly to extend it still further: one might wish to construe all empirical hypotheses without exception in the formalist fashion. From this point of view, all hypotheses would be regarded merely as more or less useful inscriptions, not as conjectures literally to be believed or disbelieved or in favor of which or against which evidence need be marshaled. This extreme view would allow us to dispense altogether with nondemonstrative argument; not even a principle of induction would be required, for even inductive generalizations would be construed just as useful strings of marks, not as statements that are true or false and that need to be confirmed or disconfirmed. This most ruthless form of the formalist view seldom is explicitly adopted, though there have been philosophers who seemed inclined toward it.[4]

In any case, formalism clearly deserves careful attention, whether the formalist view be applied only to hypotheses in which occur theoretical predicates, whether it be applied to hypotheses about the existence of unobserved entities of more familiar sorts, or whether it be ruthlessly extended to all hypotheses without exception. Let us first consider it in connection with theoretical hypotheses containing predicates not observational predicates, for here the formalist view appears to be especially attractive.

II

Among recent presentations of the formalist view of theoretical hypotheses, Braithwaite's has been the most copious;[5] but Schlick,[6] Carnap,[7] Hempel,[8] and others have expressed themselves in terms which agree to a considerable extent. Whatever their differences of emphasis, those who adopt this general viewpoint agree that the using of scientific theories is to be regarded as a game played according to rules, whether it be played just with marks or with other equipment as well. The point of the game is to derive from certain initial configurations of marks (or other equipment) certain other configurations, adhering step by step to set rules of play. Now, the configurations of

---

[4] For example, F. P. Ramsey, "Theories" and "General Propositions and Causality," in *The Foundations of Mathematics and Other Logical Essays* (London, 1931).

[5] Braithwaite, *op. cit.*

[6] Moritz Schlick, *Philosophy of Nature* (New York, 1949), p. 23.

[7] Rudolf Carnap, *Foundations of Logic and Mathematics* (Chicago, 1939).

[8] C. G. Hempel, *Fundamentals of Concept Formation in the Empirical Sciences* (Chicago, 1952).

marks (or of diagrams, dials, meter readings, and the like) which appear at various stages of the play are of two sorts; some are theoretical configurations which are not to be interpreted as expressing anything literally true or false; others are observational configurations which may be interpreted as expressing statements of a kind that admit of being verified, or at any rate inductively confirmed, by observational evidence. We shall be primarily interested in cases in which the difference between theoretical and observational formulas lies in the predicates they contain—theoretical formulas being ones that contain non-observational predicates while observational formulas contain none but observational predicates. Now, when we use a scientific theory, the initial and intermediate configurations usually will be theoretical ones, according to this view; but the final configurations derived at the end of a round of play usually will be observational ones. Thus, a scientific theory may contain formulas in which occur the marks 'atom' and 'electron,' these formulas not admitting of being inductively confirmed. But the theory will be a good one, according to this view, if from it we can derive observational formulas (predictions about meter readings or the like) which we can verify or confirm. The theory is of scientific value just insofar as it does enable us from a number of initial theoretical formulas fruitfully and systematically to derive a large number of statements that we can have reason to believe are true.

This formalist view is pragmatic in its intent. It aims to use theoretical hypotheses for deriving predictions about observable facts, thus getting the "cash value" out of them, yet without accepting them as literal statements that would need to be confirmed. But there are two somewhat different attitudes which the formalist may adopt in regard to these theoretical formulas. On the one hand, he may be thoroughly cold-blooded and may insist that they really have no significance whatever, being nothing but devices for use in the deriving of observational statements. According to this cold-blooded view, the manipulating of formulas which a scientist may perform in order to obtain an observational prediction does not differ in any essential logical respect from the incantations that a witch doctor may find it helpful to recite as a preliminary to the predictions that he perhaps makes. If the scientist generally obtains predictions which we find to be correct, we shall consider his technique good, and we shall trust his further predictions. Similarly, if a witch doctor of the better sort were to make predictions that generally were verified, then we should equally consider his technique good, and we should be inclined to trust his future predictions. The manipulating of inscriptions or of incantations as a result of which the predictions are arrived at will be equally inscrutable in either case. Just as no one would claim to understand or to believe the incantations of the witch doctor (which are not statements), so no one would be entitled to claim to understand or to believe the theoretical formulas of the scientist (which are not statements either, according to this view).

On the other hand, many formalists relent and adopt a less cold-blooded view. Formalists who do this realize how odd it sounds to say that we cannot understand or believe any of the theoretical hypotheses which science propounds about unobserved entities. Many people seem to think themselves able to understand and to believe these hypotheses; it goes against the grain to assert that they are deluded in this. Therefore, by a comfortable compromise, these formalists allow that the theoretical formulas involved possess an indirect significance in virtue of the fact that predictions are obtainable from them, and these theoretical formulas can in some sense be understood and believed; nevertheless, it is held that these formulas are not so significant as to require confirmation.

Induction and Hypothesis

III

Let us pause to ask whether this less cold-blooded formalist view about the significance of theoretical formulas really is as legitimate as the more cold-blooded one. The thinking of those who adopt this less cold-blooded formalist view seems to be colored by the semantical doctrine that expressions can be "implicitly defined," can gradually obtain meaning, through their indirect connection with expressions that do possess meaning. Thus Braithwaite tells us that the theoretical expressions in the formalist game derive some surrogate significance out of the fact that they can be used in the game, that they can be manipulated so as to yield observational statements. We know how to operate with the theoretical expressions; Braithwaite therefore thinks that we may be said to understand them. These theoretical expressions "are not understood as having any meaning apart from their place in such a calculus," yet "we give direct meanings to those formulae of the calculus which we take to represent propositions about observable entities; we give indirect meanings to the other formulae as representing propositions in a deductive system in which the observable propositions are conclusions." [9] It is evidently supposed that by deriving more and more statements from a formula that is not a statement we are enabled to attain a better and better understanding of the previously senseless expressions in that formula, so that these expressions gradually "take on" empirical significance, and the formula becomes more and more nearly a statement.

This is a rather strange semantical doctrine. Those who speak in this way actually seem to be assuming that any derivation of statements from formulas which are not statements will help give meaning to such marks as occur uninterpreted in those formulas. But surely so far as logic is concerned significance is

---

[9] Braithwaite, *op. cit.*, p. 51.

not a garment with which naked marks can gradually come to be clothed. From the logical point of view, expressions either have or have not got semantical rules governing their interpretation. If the expressions do have significance, it should be possible to say what their significance is. But if they have not, it is no good talking about some suppositious process of their "coming to have" it. Psychology (even good psychology) cannot be a substitute for logic, and the mingling of psychological with logical issues is bound to cause confusion. So long as we are adopting a logical point of view, we cannot countenance expressions that go about acquiring or altering their significance without giving notice.

From the way they talk, one might gather that some of the philosophers who speak in this way really have in the backs of their minds the thought that a logically sufficient condition for a predicate's having empirical significance is that it should occur in a formula from which significant statements are derivable. But this is not very plausible; such a latitudinarian criterion of empirical significance would not exclude anything—scientific theories would not be distinguished from metaphysical discourse or even from sheer babbling. According to such a criterion, supposing it be granted that 'Jones is six feet tall' is a significant sentence, it would follow that the formula 'Jones is a tove and all toves are six feet tall' must be significant also merely because the former statement can be derived from it. But the predicate 'tove' and the formula in which it occurs cannot be said to gain empirical significance from the mere fact that a significant sentence is derivable in this way; for we are left absolutely none the wiser as to the sense in which the predicate 'tove' is to be used. In spite of this derivation the word 'tove' remains just as much a nonsense syllable as ever; and to use the nonsense syllable in a million different syllogisms could leave it as much a nonsense syllable as before.

Induction and Hypothesis

A considerably more sophisticated view of the empirical significance of theoretical terms has, however, been adopted by some other philosophers, notably by Carnap. He believes that scientific hypotheses ought to contain predicates which are not definable by means of observational predicates; he wants to say that these theoretical predicates are only "partially interpreted" and that such significance as they possess they have in virtue of their belonging to a system from which observational statements are derivable. But he feels the need to maintain a distinction between scientific theories, which deserve to be called empirically significant, and other kinds of discourse, such as metaphysics, which do not deserve that. Various philosophers have made attempts to mark this distinction, and Hempel has lucidly chronicled their lack of success.[10] Carnap, however, has recently proposed a new and more careful criterion based upon the thought that a theoretical predicate should rate as empirically significant in a system if it makes a difference with regard to the predicting of some empirically significant fact.[11] Putting the matter in linguistic terms, suppose we have a language (possessing a suitable logic) in which can be formulated a set of theoretical postulates T (none containing any observational predicate) and also a set of postulates C which serve as "co-ordinating rules" (each of them contains both theoretical and observational predicates). Carnap's criterion of empirical significance for terms then can be paraphrased as follows:

---

[10] C. G. Hempel, "Problems and Changes in the Empiricist Criterion of Meaning," *Revue internationale de philosophie*, IV (1950), 41–63, and "The Concept of Cognitive Significance: A Reconsideration," *Proceedings of the American Academy of Arts and Sciences*, LXXX (1951), 61–77.

[11] Rudolf Carnap, "The Methodological Character of Theoretical Concepts," in Herbert Feigl and Michael Scriven, eds., *The Foundations of Science and the Concepts of Psychology and Psychoanalysis* (Minnesota Studies in the Philosophy of Science, vol. I; Minneapolis, 1956), pp. 38–76.

Observational predicates are empirically significant;

Any further term M in the language is empirically significant if and only if it satisfies all the following requirements:

The language contains a sentence S_M which has M as its only predicate;

The language contains a sentence S_K all of whose extralogical terms can be shown to be empirically significant;

The conjunction of S_M with S_K and T and C is not logically false;

The language contains a sentence S_0 which has observational but no other extralogical terms;

S_0 is implied by the conjunction of S_M with S_K and T and C, but is not implied by the conjunction of S_K with T and C.

If all the extralogical terms occurring in T and C satisfy this criterion, then the theoretical system whose postulates the items of T and C are will be an empirically significant system.

If this criterion of Carnap's provided a satisfactory distinction between proper and improper use of theoretical terms in science, then this would make it more plausible to suppose that properly used theoretical terms should be called empirically significant even if only partially interpreted. Unfortunately the criterion does not seem to be able really to establish the kind of distinction that we should wish to find between significant and nonsignificant uses of terms.

Let us consider an example. Suppose that someone has constructed a metaphysical theory about the Absolute. This theory possesses a set of postulates T in none of which occurs any observational term at all: 'There is one and only one Absolute,' 'The Absolute and it alone is such that of any two distinct things one manifests it more than does the other,' and the like.

The theory also possesses another set of postulates C, each of which contains observational terms as well as metaphysical terms: 'If one nation defeats another in war, then the former manifests the Absolute more than does the latter,' 'The Absolute is not an animal but animals manifest the Absolute more than do vegetables,' and so on. Quite an elaborate theory of the Absolute might be built up, yet nevertheless all these postulates tell us nothing at all about how to recognize the Absolute, nor do they enable us to make any new predictions about observed phenomena; they represent purely metaphysical speculations. Surely this is exactly the sort of theory which the empiricist would wish to dismiss as lacking empirical significance. However, the inventor of this theory may dislike being ostracized, and perhaps he will wish to legitimize his theory in the eyes of empiricists, modifying it slightly so as to satisfy Carnap's requirements. One way to do this would be to augment the postulates C by two new postulates such as: 'If a penny is pocketed, then it is absolute if and only if it instantly turns into a dollar,' and 'If x, y, z, are dogs, then x manifests z more than does y if and only if x barks louder at z than does y.' These two new postulates do not inconvenience the metaphysician: he wants to talk about manifestations only of the Absolute (which is not an animal), so he does not care what sense the term 'manifests' is assigned with regard manifestations of a dog; and he is quite willing to allow that pennies (none of which do turn into dollars) are not absolute; moreover, he may be quite content to say that if *per impossibile* a pocketed penny were to turn into a dollar it would deserve to be called absolute, since only the Absolute could perform such a miraculous feat. With these two new postulates added, it now is possible to show that the system satisfies Carnap's criterion of empirical significance. Consider first the predicate 'manifests.' Choose S_M as 'Fido manifests Rover more than does Bonzo,' choose S_K as 'Fido, Rover, and

Bonzo are dogs,' and S_O as 'Fido barks louder at Rover than does Bonzo.' Here the conjunction of S_M with S_K and with T and C implies S_O, but the conjunction of S_K with T and C does not. Thus, according to the criterion, 'manifests' counts as an empirically significant predicate in this system. Now consider the predicate 'absolute' (by means of which the singular term 'the Absolute' can be defined). Take S_M as 'This penny is not absolute,' take S_K as 'This penny is pocketed,' and take S_O as 'This penny does not instantly turn into a dollar.' Now putting these statements together in the manner that is called for by Carnap's criterion of empirical significance, we see that the conjunction of statements S_M, S_K, T and C implies S_O, while the conjunction of S_K, T, C does not. Thus 'absolute' satisfies the criterion of empirical significance also. And since all the remaining predicates of this theory are observational predicates, the system as a whole must rate as an empirically significant system, according to Carnap's criterion. The inventor of the theory, having complied with the requirements, now can happily return to his speculations about the Absolute and its manifestations.

This example suffices to suggest how a body of discourse that is thoroughly metaphysical can with very inessential modifications be made to satisfy the criterion of empirical significance which Carnap proposes. All that is necessary is to invoke a new postulate for each of the various special predicatives involved, choosing these postulates so that the uses of these special predicates will be partially determined in ways which are consonant with but irrelevant to the intention of the metaphysical speculations. The fact that this can rather easily be done would seem to cast doubt upon the adequacy of Carnap's criterion of empirical significance.

Reflection would seem to show that one scarcely could expect a criterion of empirical significance to be free from difficulties like this, so long as the criterion attempts to rate partially interpreted terms as significant. So long as one insists upon regarding terms as empirically significant to which no explicit sense has been assigned, it would seem that one cannot be in a position to exclude metaphysical discourse. But this is highly unsatisfactory, for we do want to preserve if we can the distinction between science and nonempirical metaphysics. Might it not be better to abandon the view that theoretical terms in some way derive surrogate significance out of their indirect connection with observations? Then we could fall back upon the more cold-blooded view that theoretical formulas, though perhaps useful, are strictly speaking just mumbo-jumbo. We could adhere to the view that in logical argument the initial terms must have significance to begin with if significance is to be involved at all and that the argument itself cannot give it to them. We could maintain that syntactical rules, which permit manipulations of formulas, are no substitute for semantical rules, which stipulate truth-conditions for those formulas.

IV

If we cannot accept the view that theoretical formulas possess some kind of partial or surrogate significance merely in virtue of their place in a system, then we shall fall back upon the cold-blooded formalist view that they have no significance. But is this a plausible view? In certain respects this formalist view seems paradoxical. Some people will feel that the picture of science which it offers us is intuitively dissatisfying. The source of this discontent is that the formalist seems to be claiming that scientific theories are not true or false in any literal sense; yet at the same time he seems to be claiming that they are essential to scientific explanation. He seems to be working both sides of the street.

If the theories in question could be regarded as abbreviations in the reductionist's sense, then it would not seem paradoxical to deny that they make literal assertions; for in that case to state a scientific theory would at least be to make some assertion, though it be an assertion which might require to be translated into other terms whenever we wished to speak strictly and officially. But such translations, as we saw, are not possible in general; and what the formalist now wants us to believe is that science (and also much of everyday knowledge), while seeming to give us information about how the world is, instead really consists largely of formulas so lacking in meaning as neither to be true nor false at all. With one breath he grants to us that theories are indispensable to science; with the next breath he tells us that they are merely convenient dodges, devices useful in the deriving of observational statements. Is science then to consist of a repertory of tricks instead of a body of knowledge? Science then becomes an art rather than a science: its theories may be polished or slovenly, handy or futile, but not true or false. Yet we started off with the supposition that science was to consist of a body of statements expressing our knowledge about the empirical world, these statements to be confirmed by the evidence which experience provides. Now the formalist tells us that we ought to admit into our science, as irreducible and official constituents, inscriptions which are mere tools. But uninterpreted marks, even useful ones, cannot constitute knowledge.

Especially it seems strange to claim that uninterpreted marks can play any vital role in scientific explanations. If the formalist view were correct, scientific explanation would be an odd procedure. Instead of explaining a fact by showing how it follows from hypotheses that have been confirmed by previous evidence, what the formalist does is to show that in his game with marks he can start from uninterpreted theoretical inscriptions and can derive a statement expressing the fact to be explained. This derivation he would call an explanation. For instance, if we start with the formulas 'Jones had an unhappy childhood,' 'Anyone unhappy in childhood is a tove,' and 'All toves are neurotic,' then we can derive the consequent 'Jones is neurotic.' And presumably this derivation would count as a scientific explanation, according to formalist principles. But it is odd to contend that this explains Jones's neurosis or that the evidence that Jones is neurotic could help us to establish the theory that anyone unhappy in childhood is a tove and that all toves are neurotic. This use of ostensible discourse about inscrutable entities seems pointless; yet the formalist would have to think of it as a paradigm of scientific reasoning.

The formalist would have us believe that by juggling with these inscriptions it is possible to explain observable facts; but our intuitive feelings as to what an explanation ought to be are upset by this contention. Playing some elaborate game of noughts-and-crosses will not suffice to explain the clicking of a Geiger counter, nor is a query as to the cause of a disease properly answered by some adroit incantation of nonsense syllables. This is not what we mean by explanation. Empiricist philosophers sometimes have accused Scholastic philosophers of trying to offer in place of explanations mere complicated and dextrous manipulations of senseless expressions. It seems incongruous to see empiricists themselves now explicitly embracing this very practice on a more sweeping scale.

V

The formalist maintains that theoretical formulas, although they are not to be regarded as true statements, nevertheless have an important role to play in science. Their role is to enable us to derive new observational predictions and to enable us to explain observations that have already been made. The preceding section suggested that this view is intuitively dissatisfying; but there is a further and sharper difficulty as well.

Formalism

Suppose that we have available some definite supply of observational evidence, and suppose that we are trying to decide what scientific theory we ought to base upon it. What theory, among all those that might be invented, shall we choose as best on the basis of this evidence? The formalist will tell us that we should choose as theory an uninterpreted system that enables us to derive inscriptions which can be interpreted as observational statements expressing our evidence (or perhaps as statements confirmed by our evidence). But unfortunately this is not much of a criterion; it does not enable us to decide what theory we ought to adopt. For what can be derived in one game with marks can equally well be derived in any number of other very different games. Any number of "calculi" might be constructed, each of which makes it possible to derive the specified set of observational statements; some of these calculi would be trivial, some cumbrous, a few interesting. But we cannot rest satisfied with an unlimited number of very different yet equally good theories: in science we must choose among these theories, we must be able to say that some of them are better than others on the basis of the evidence.

The difficulty involved here emerges more clearly if we consider some specific theory in science and if we suppose that all the actual evidence upon which this theory is based can be formulated in observational statements O_1, O_2, . . . O_n. How would the formalist explain the sense in which this theory is based upon the evidence? In order that the theory perform the role which the formalist assigns to it, what is necessary is that the theory should constitute a system having rules and initial formulas from which can be derived some set of statements S_1, S_2, . . . S_i . . . some of which are statements already well confirmed or verified, others perhaps being as yet untested. But given this evidence that we do have, how should we decide what system to adopt? There are any number of systems fulfilling the requirement mentioned. For instance, one such system would be that which takes as its initial formulas just O_1, O_2, . . . O_n and which has no rule other than that any initial formula may be obtained at will. In this trivial system are derivable the statements O_1, O_2, . . . O_n, each of which is conclusively verified. It would seem, on formalist principles, that this "theory" should be reckoned at least as good as any other. Why bother to construct a more elaborate scientific theory than this?

Now, it may be objected that this utterly trivial system does not constitute a very good scientific theory because it yields no new predictions, no tests for the future; it enables us to derive only the evidence which has already been verified. To this one might reply that the goodness of a theory ought to be judged by considering whether it fits the evidence we actually have, not whether it fits evidence that we have not got. However, if predictions are desired, let them be added to our trivial system: let O_{n+1}, O_{n+2}, . . . O_{n+m} be whatever predictions you like, and let our system be modified so as to have as its initial formulas O_1, O_2, . . . O_{n+m}. We now have a system in which all the evidence can be derived plus the desired predictions; is not this system just as good a theory as any? Indeed, our trivial system, on account of the simplicity of its rule, is much easier to manipulate than are most others, and (so it would seem) ought to be preferred.

What is the point of this? Simply that the formalist account of what a theory does has the consequence that it would be pointless for anyone to waste time in constructing a scientific theory that was not trivial; for the job that the formalist assigns to a theory is a job that can quite conveniently be done by a theory of painful triviality. If the formalist were right, scientists would all be misguided when they spend their time constructing subtle and intricate theories. And this appears to be an unreasonable consequence of the formalist view.

Moreover, the trivial system just considered enables us to derive the desired observational statements without employing any theoretical expression at all. In the initial formulas of this little system there do not occur any predicates that are not observational predicates, nor in the system do there occur any formulas at all that are not couched wholly in observational language. Thus there is no need in this case of juggling uninterpreted inscriptions; all the marks in this system do admit of being interpreted in a straightforward way. This example may lead us to wonder whether in general it will always be possible to avoid using theoretical expressions if we wish.

In fact, it always is possible to do without theoretical expressions. Given any system whatever from which a certain body of observational statements are derivable, it always is possible to construct another system which is such as to contain no theoretical expressions at all and which is such that from it are derivable all and only those observational statements that were derivable from the first system. This important fact has been established by Craig.[12]

This fact is important because it shows that there cannot be any logical advantage to be gained by introducing theoretical expressions into a scientific system. By introducing theoretical expressions one cannot obtain a system which is richer than would otherwise be possible with regard to the observational statements derivable from it (richer in its "cash value") or more economical than would otherwise be possible with regard to its array of primitive terms. Whatever degree of richness and economy may be desired, these always can be secured at least as well through the adoption of a system containing no theoretical expressions whatever. A system containing theoretical expressions might be handier or more compendious, in some

---

[12] William Craig, "Replacement of Auxiliary Expressions," *Philosophical Review*, LXV (1956), 38–55.

cases; but there seem to be no logical grounds on the basis of which it ever could claim to be in any sense better supported by the evidence. It is true that if we take a system containing theoretical terms and transform it by Craig's method into a new system containing none but observational primitives, this new system may need to have more than a finite number of postulates; but this is no defect, for its postulates will nevertheless admit of being effectively characterized—and some perfectly respectable systems (for example, Quine's *Mathematical Logic*) are so codified as to allow for an infinite number of postulates.

In spite of this, the formalist may maintain that the system containing theoretical expressions is to be preferred because it is somehow more convenient or more fruitful than would be a system from which theoretical expressions had been excluded.[13] But it is far from easy to know what to make of these vague terms 'convenience' and 'fruitfulness.' Is one system more convenient than another if its rules and initial formulas can be written down with fewer strokes of the pen or if derivations within it are usually shorter in length? Or should we consider instead the quantity of muscular or of intellectual effort required in using the system? Now, of course there are various respects in which some theories are more convenient, more suggestive, or more fruitful than others; and some of these respects could be clarified. Certainly it is proper for scientists to use convenient and fruitful theories. But it would seem that convenience and fruitfulness are merely heuristic considerations and do not provide us with any criterion for deciding what scientific theories really are best supported by the available evidence. Scientists surely do believe that some theories are well supported by the evidence while others are not; but the formal-

---

[13] See C. G. Hempel, "Implications of Carnap's work for the Philosophy of Science," in *The Philosophy of Rudolf Carnap* (Library of Living Philosophers), to be published.

ist's appeal to the extremely vague notions of convenience and fruitfulness does not account for this belief. The formalist appears unable to give any convincing reason why one scientific theory ever should be best or why we should adopt any nontrivial theories at all.

VI

So far we have been considering the formalist view of theoretical hypotheses, those which contain primitive predicates that are not observational predicates. Hypotheses of this character cannot be confirmed inductively, nor do they appear to make any logical contribution at all to systems into which they may be incorporated; whatever quantity of observational consequents and whatever degree of economy of primitives may be desired, these always can be obtained without the use of any nonobservational predicates at all. In the light of this, the formalist's attempted justification of the use of such hypotheses seems unconvincing. Moreover, it begins to look doubtful that such hypotheses ought to be regarded as playing any fundamental role at all in empirical knowledge. Might it not be better to insist that (at least when we speak strictly) all predicates appearing in our empirical hypotheses be observational predicates or definable by means of observational predicates? This would free us from the curious perplexities that surround theoretical hypotheses. We should then be obliged to claim that such terms as 'electron,' 'atom,' and 'gene' should be definable by means of observational predicates, if they are more than heuristic devices, if they occur in hypotheses which we regard as established or supported by scientific investigation. Nor is it clear that this would be an absurd claim, for formalists generally have leaped far too hastily to the conclusion that such definitions are impossible—with ingenuity they might well be devised satisfactorily.

Induction and Hypothesis

But even if this line is followed with regard to hypotheses ostensibly containing theoretical terms, we shall still have to consider transcendent hypotheses which are not theoretical—hypotheses containing only observational predicates yet not confirmable by induction because they imply the existence of unobserved things. A formalist might claim that these hypotheses, since they cannot be confirmed by induction, ought to be construed not as statements but merely as inscriptions which are useful in enabling us to derive predictions about observed phenomena. Or he might even go so far as to claim that all hypotheses ought to be construed in this formalist fashion. What can we say about this application of the formalist view? As might be expected from what has already been said, several difficulties arise.

In the first place, this formalist view would assert that sentences purporting to refer to unobserved objects are to be construed as uninterpreted marks and therefore cannot be true or false; and this seems odd. It would be more plausible to suppose that sentences of acceptable logical forms containing none but observational predicates ought to count as genuine statements, indeed as empirically significant ones. Even in our everyday thinking we employ innumerable hypotheses about things which we do not directly observe; and it would be strange to deny that these hypotheses can be true or false or to claim that we do not need to marshal evidence in their favor.

In the second place, this conception of the role to be played by these hypotheses seems unintuitive in that again we are presented with a strange picture of the nature of explanation. If one observes a compass needle being deflected near a locked desk drawer, it would be natural to explain this by saying that there is a magnet inside the drawer; but it seems distinctly unnatural to insist that the explanation should consist merely in the fact that from the inscription 'A magnet is in the drawer' the rules of some game or other allow us to derive the statement 'The compass needle is deflected.' The explanatory statement cannot be an uninterpreted inscription, for we shall accept the explanation as an adequate one only if we believe the explanatory statement true.

Finally, there remains a difficulty about the difference between good hypotheses and poor ones. In everyday thinking as in science we assume that there is an important difference between hypotheses well supported by the evidence and those which are not. But the formalist view of hypotheses makes it impossible to understand this difference. If the only job that hypotheses have is to assist us in a game of deriving predictions about things that are directly observed, then there is no clear way of deciding what hypotheses to adopt; innumerable utterly different games could be constructed, any of which might allow us to derive the desired predictions. The formalist view gives us no guidance concerning how we are to choose among these.

Thus we seem obliged to conclude that the formalist view is no more satisfactory for hypotheses couched in observational language than it was for those supposedly containing theoretical terms. The problem of explaining how our hypotheses are to be confirmed cannot be avoided by means of the supposition that they are uninterpreted inscriptions.

Eight

The Method of Hypothesis

I

INDUCTION seemed unable to account for the confirming of hypotheses implying the existence of unobserved things, and so we have been examining ways in which such hypotheses supposedly might be dispensed with. Reductionism sought to construe all such hypotheses as mere abbreviations for more cumbrous hypotheses about observed entities; formalism sought to construe them as partially interpreted or as uninterpreted inscriptions which admit of being manipulated so as to yield predictions concerning observed entities. But we have seen that both reductionism and formalism lead to incoherency, and it appeared that neither was acceptable. It looks as though there is no way of dispensing with these hypotheses which imply the existence of unobserved things; we need to retain them as essential elements in our empirical knowledge.

If it be granted that this is so, then we must seek a principle of nondemonstrative inference which will warrant the support that such hypotheses need from observational evidence. Mere induction did not suffice, and an inductivist could account for the confirming of these hypotheses only if he were to adopt some strong factual assumptions about the nature of the world (assumptions which, like Russell's postulates, would be difficult to state and impossible to justify). It appears that we should do better to seek some quite different, noninductive mode of inference.

Examination of the literature on the subject reveals only one other general sort of method that has been proposed as a fundamental alternative to induction. Though seldom precisely stated, this is a method which may be traced back to Plato's practice of "saving the appearances" by constructing empirical theories which, though without inductive warrant, did serve to explain observed phenomena.[1] A few of its proponents have suggested that this method is not based upon induction but that it is at least as fundamental as induction, perhaps more so. Broad has called it "the hypothetical method," and it also has been referred to more pretentiously as "the hypothetico-deductive method."[2] The latter phrase sometimes has a connotation of formalism, and the former phrase may convey the misleading suggestion that this method is solely concerned with hypothetical statements; so perhaps it will be best if we simply speak of "the method of hypothesis."

To begin with, let us give only a very rough characterization of this method, since that is all that can be got out of the accounts usually offered of it. Roughly, then, the method of hypothesis consists in deducing consequents from a hypothesis and in verifying them; the hypothesis is regarded as confirmed if some consequents are verified and none is falsified; and one

---

[1] For instance, throughout much of the *Timaeus*.

[2] C. D. Broad, "On the Relation between Induction and Probability," *Mind*, XXVII (1918), 390.

hypothesis is regarded as better confirmed than is another if more consequents of the former than of the latter have been verified and none falsified. In an earlier chapter we saw that obscurities can lurk in a rough statement like this, and we saw that this sort of method of nondemonstrative inference cannot merely be based upon induction. However, it may be hoped that the obscurities can be clarified and that the method may deserve to be adopted not as derivative from induction but as a fundamental method, acceptable in its own right.

II

The method of hypothesis has some intuitive appeal, for it accords with a certain fairly common view of the nature of scientific reasoning. According to this view, the task of science is to explain phenomena, and a scientific hypothesis strengthens its credibility just to the extent that it does succeed in explaining observed phenomena. What grounds have we for believing the theories of physics? We are tempted to reply, well, they explain the observed facts, don't they? And similarly for more commonplace examples: what grounds are there for believing the cat to be behind the sofa? Well, if it were, that would explain the mewing and the protruding tail. This view about nondemonstrative inference is most boldly stated by Whewell when he writes that the deducibility of the premises of a nondemonstrative argument from the conclusion constitutes "the criterion of inductive truth" (that is, the criterion of the soundness of the nondemonstrative argument).[3]

This common-sense way of putting the matter may have seemed satisfactory to philosophers of earlier generations. But nowadays one cannot rest satisfied with such rough formulations, in particular because Hempel's illuminating discussions

---

[3] William Whewell, *Novum Organon Renovatum*, 3d ed. (London, 1855), pp. 114–115.

have shown the kinds of unsuspected logical difficulties to which they are subject.[4] If we look critically at this loose and common-sense way of formulating the method of hypothesis, we can see that it suffers from two especial shortcomings.

In the first place, the uncritical proponent of the method of hypothesis is claiming that any hypothesis is confirmed if and only if consequents deducible from it are verified; but this would entail that many hypotheses which we set store by cannot be confirmed at all in a direct and natural way. Hypotheses universal in form would fall under this ban, for instance. Only observational statements are verifiable, and an observational statement needs to be existential in form; but from a universal statement no existential statement can be deduced. Thus no universal statement (considered in isolation) can have any verifiable consequent, and thus no universal statement can be directly confirmed. This would be a serious defect of the method of hypothesis.

In the second place, this loose formulation of the method of hypothesis would have the consequence that many hypotheses which do not deserve to be confirmed could gain confirmation in an indirect way. For instance, let *H* be any confirmable hypothesis whatever; now consider the hypothesis which is the conjunction of *H* with 'The Over-arching Self differentiates itself into self and nonself whilst yet remaining one with itself.' This conjunctive hypothesis must be at least as well confirmed as is *H*, according to this loose formulation of the method of hypothesis, since from this conjunctive hypothesis are deducible

---

[4] C. G. Hempel, "A Purely Syntactical Definition of Confirmation," *Journal of Symbolic Logic*, VIII (1943), 122–43, and "Studies in the Logic of Confirmation," *Mind*, LIV (1945), 1–26, 97–121. These papers also contain a valuable positive theory of confirmation, which, however, applies primarily to languages containing none but one-placed predicates. A discussion of parts of these papers is to be found in Rudolf Carnap, *Logical Foundations of Probability* (Chicago, 1950), secs. 87 and 88.

all and only those verifiable consequents that are deducible from *H* alone. Moreover, the metaphysical statement, since it is implied by the conjunctive hypothesis, can be no less well confirmed than is that hypothesis. Therefore, by this indirect bit of reasoning, we conclude that the metaphysical statement must be at least as well confirmed as is *H*, whatever *H* may be. And this is not a satisfactory result.

These two simple objections suffice to show that the proposed method of hypothesis is not at all acceptable when stated in this uncritical fashion. Hypotheses which we want to confirm cannot be confirmed in any straightforward way, while outrageous statements do receive confirmation in an indirect way. If the method of hypothesis is to be at all plausible, it must somehow be restated. It must be reformulated so as to be at once generous enough to allow the confirming of hypotheses which do provide scientific knowledge yet also restrictive enough to prevent the confirming of extravagant nonempirical statements. A satisfactory criterion must permit the confirming of good empirical hypotheses and must enforce the disconfirming of bad ones.

III

The formulation of the method of hypothesis so far discussed has been a casual one lacking in exactness. Popper, however, constructed a theory of a somewhat more detailed and careful sort.[5] His theory represents an unusual variant of the method of hypothesis, for he argued that a hypothesis is confirmed not through the verification of its own consequents but rather through the falsification of some of its rival hypotheses (which, to be sure, usually ensues upon the verification of some of its own consequents). In his view, this method of falsification is

---

[5] Karl Popper, *Logik der Forschung* (Vienna, 1935).

The Method of Hypothesis

fundamental to all nondemonstrative inference and is in no way dependent upon induction.

To begin with, Popper asserts that a hypothesis is to count as empirically testable if and only if some observational statement can be formulated which contradicts the hypothesis. Thus the generalization 'All ravens are black' will be admissible into science as a testable hypothesis provided the statement 'There is a raven that is not black' is a statement of the sort which admits of being verified by direct observations. This may not be too satisfactory as a criterion of admissibility for scientific hypotheses;[6] indeed, at a later stage of his exposition Popper does countenance other hypotheses which cannot be falsified directly (for instance, hypotheses about "points of condensation" in certain endless sequences of events—hypotheses which figure in Popper's version of the frequency theory of probability). But the main thing of interest to us here is the use Popper makes of the notion of falsifiability in order to provide a criterion for distinguishing among the degrees of confirmation which empirical hypotheses possess.

We need to have a criterion by means of which to judge whether one hypothesis is better confirmed than is another, on the basis of given observational evidence. What Popper suggests is that when we have two hypotheses neither of which is falsified by the evidence we ought to regard the more falsifiable hypothesis as the better confirmed. The better confirmed hypothesis is to be the one which "says more," which "forbids more," which is the "more testable." Thus to say that hypothesis H_1 is better confirmed by evidence E than is hypothesis H_2 (where H_1 and H_2 are each compatible with E) is to say that H_1 is capable of being falsified by a greater number of possible observational statements than is H_2. According to Popper's

---

[6] C. G. Hempel, "Problems and Changes in the Empiricist Criterion of Meaning," *Revue internationale de philosophie*, IV (1950), 41–63.

scheme it will often (perhaps always) be the case that an empirical hypothesis will be falsifiable by innumerably many possible observational statements. Where this is so, we need some explanation of what it means to say that one hypothesis is falsifiable by "more" observational statements than is another. Popper gives two partial explanations of this: he says that one hypothesis is falsifiable by "more" observational statements than is another provided the former is falsifiable by every observational statement which can falsify the latter plus some others as well; and he offers a doctrine of "dimensional comparison" in terms of which certain mathematically complex hypotheses count as less falsifiable than certain mathematically simpler ones. This notion of degrees of falsifiability thus enables Popper to explain his claim that the preferable hypothesis is the one which leaves itself more open to falsification and yet manages to survive.

Popper's view does have the signal virtue of offering us a really noninductive theory of nondemonstrative inference. But is his view an acceptable one? It may be objected at once that his view allows us to compare degrees of confirmation in too few cases; for often we should wish to assign different degrees of confirmation to two hypotheses neither of which has the "falsification possibilities" of the other as a proper subclass of its own and neither of which is mathematical. But waiving this, there are two other serious objections which seem to apply to Popper's view, objections not dissimilar to those already levied against the uncritical formulation of the method of hypothesis.

In the first place, Popper's criterion of confirmation seems to allow the confirming of nonempirical hypotheses, because it requires us to assign unduly high degrees of confirmation to certain kinds of compound hypotheses. For example, suppose p is a falsifiable statement; then any statement of the form '$p.q$' or '$q.q \supset p$' or '$q.q \equiv p$' will be at least as falsifiable and hence will have to count as at least as well confirmed as is p itself. More concretely, if 'All swans are white' is a falsifiable statement, then 'All swans are white, and the World Spirit informs all activity' must be just as falsifiable a statement, since any observational statement which contradicts the former must contradict the latter also; hence, whatever the evidence may be, the latter statement can be no less well confirmed than the former. Similarly, 'The entelechy causes growth, and the entelechy causes growth only if all swans are white' and 'The soul is the form of the body, and the soul is the form of the body if and only if all swans are white' must be accounted just as empirical, just as falsifiable, and just as well confirmed whatever may be the evidence, as is 'All swans are white.' This result is curious; and it becomes more curious when we recall that 'The World Spirit informs all activity, and all swans are white' implies 'The World Spirit informs all activity'; the former statement cannot be true unless the latter is so, and consequently the degree of confirmation of the latter can never be less than that of the former. According to Popper's criterion, therefore, we are obliged to conclude that 'The World Spirit informs all activity' must be a hypothesis at least as well confirmed, whatever the evidence may be, as is the hypothesis 'All swans are white.'

Further examples arise if we take account of quantificational as well as of truth-functional forms. For example, suppose that 'is a raven' and 'is black' are observational predicates such that 'There is a raven which is not black' is a possible observational statement. Now consider a hypothesis such as 'All ravens are black': thus must be a falsifiable hypothesis. But the hypotheses 'All ravens are reincarnations of our ancestors, and all reincarnations of our ancestors are black' and also 'No ravens are reincarnations of our ancestors, and whatever is not a reincarnation of an ancestor is black' are hypotheses which are equally falsifiable and will have to be regarded as just as well confirmed, whatever the evidence may be, as is 'All ravens are black.' Moreover, 'All ravens are reincarnations of our ancestors' and 'No ravens are reincarnations of our ancestors,' since they are respectively implied by the compound hypotheses just mentioned, cannot be less well confirmed than they. All this would make biology rather difficult.

So far we have been considering one objection against Popper's view of confirmation; but a second objection can be raised without appeal to metaphysical hypotheses, for Popper's scheme seems to have a defect in regard to the way in which it rates the relative degrees of confirmation of some perfectly good empirical hypotheses. Popper holds that of any two hypotheses the more falsifiable (so long as it has not actually been falsified) is always to be regarded as the better confirmed, and he goes so far as to criticize Keynes for having held that in some cases the hypothesis of lesser falsifiability may be better confirmed. Keynes had held that a generalization of lesser scope will be more probable than one of broader scope, other things being equal; [7] for instance, 'All men are mortal' will be more probable than 'All animals are mortal.' But Popper notes that the latter generalization is the more falsifiable, and therefore he contends that it is to be regarded as the better confirmed.[8] But surely Popper's view is not tenable here. Whenever one hypothesis implies another, as is the case in our example, the one implied must be at least as well confirmed as is the one that implies it; for the latter may be false although the former is true, but the former cannot be false and the latter true. In such a case to call the more falsifiable hypothesis better confirmed is patently to misuse this phrase.

These two difficulties that affect Popper's theory are enough

---

[7] J. M. Keynes, A *Treatise on Probability* (London, 1921), pp. 224 f.
[8] Popper, *op. cit.*, sec. 83.

to show that his scheme falls short of providing a satisfactory account of nondemonstrative inference. Nevertheless, Popper's theory is noteworthy as an attempt to work out the logic of confirmation on a noninductive basis. The great majority of other writers on this subject have adopted induction, sometimes quite unquestioningly, as the fundamental method by which hypotheses are to be confirmed. In preceding chapters we have seen some of the difficulties to which inductivism appears to lead, and in seeking to avoid these difficulties it is only natural that we should turn with interest toward the method of hypothesis. We may be disappointed to find that Popper's version of this method has flaws, but his views are suggestive and perhaps hint at ways out of the impasse of inductivism.

IV

The uncritical common-sense version of the method of hypothesis and Popper's more sophisticated variant upon it alike have appeared to be unsatisfactory, both with regard to what they permit and what they prevent. Especially troublesome is the fact that they rate a complicated and implausible hypothesis as being at least as well confirmed as is any simpler hypothesis which it implies. In concrete cases how should we try to circumvent this defect?

When faced with a concrete case in which the observed phenomena can be explained by various hypotheses, some more complex than others, we should be inclined to invoke the notion of simplicity. Other things being equal, it is the simpler hypothesis which ought to be regarded as the better confirmed. Thus, if the mewing and protruding tail can be explained both by the hypothesis that one cat is behind the sofa and also by the hypothesis that ten cats are there, then, other things being equal, it is the simpler, less prodigal hypothesis which we should prefer and which we should rate as the better confirmed. Moreover, the notion of simplicity seems to be of considerable importance in science as well as in more homely examples. In considering rival scientific theories, what will especially be weighed is their relative simplicity or parsimoniousness, both of assumptions and of fundamental concepts. The simpler theory will be regarded as better supported by the evidence, while the more complex theory, even if the evidence be equally consonant with it, will be regarded as less well supported. *Enim Natura simplex est* is a phrase of Newton's, evidently intended as a methodological maxim.

Of similar import are the Scholastic principles of which Occam made such telling use that they have since borne his name: *frustra fit per plura quod per pauciora potest fieri* and *entia non sunt multiplicanda praeter necessitatem*. These maxims are appropriate both to empirical science and to philosophical theorizing; we use them frequently, whether we are conscious of it or not. The thought involved is referred to by Sir William Hamilton as "the law of parsimony; which prohibits, without a proven necessity, the multiplication of entities, powers, principles, or causes." [9] The idea that simplicity is essential to the establishing of theories is reflected also in the Idealistic notion that in all thinking there is a nisus toward "coherent system"; [10] for "coherence" if it means anything must mean that the theories adopted are to form an integrated and economical, that is, a simple, whole.

The notion of simplicity has a certain attractiveness, yet at the same time it is also repelling in its vagueness. What can simplicity mean, if we try to construe it in some exact sense? The difficulty of giving any precise account of it has led many

---

[9] Sir William Hamilton, *Discussions on Philosophy and Education*, 2d ed. (New York, 1853), p. 580.

[10] F. H. Bradley, *Essays on Truth and Reality* (Oxford, 1914), pp. 209–212.

The Method of Hypothesis

people to conclude that the notion can have no exact sense but is merely a vague compound of pragmatic and aesthetic considerations. But perhaps this pessimistic reaction is not wholly justified.

If we do try to make something of the notion of simplicity, we shall want to say that there are three principal respects in which theories and hypotheses seem to admit of variations in simplicity. They may vary with regard to the number of individual entities which they assert to exist; they may vary with regard to the number and complexity of the independent concepts (or kinds of entities) which they involve; and they may vary in the number and complexity of the statements which they contain. Thus a theory postulating the existence of one unobserved planet is to be preferred to a theory postulating two, other things being equal; a theory explaining atomic phenomena in terms of five kinds of elementary particles is preferable to one requiring ten; and three laws of motion are better than a larger number. Of course these are no more than rules of thumb; yet they are employed in the common-sense evaluation of degrees of confirmation, and they do have some plausibility.

But these rules of thumb are ambiguous, vague, and disparate. In the first place, they are ambiguous, for in case of conflict where one rule suggests one result and another rule suggests some other result, we have no way of adjudicating between them, no means of obtaining a definite answer. For instance, if one theory postulates fewer entities but involves a smaller number of statements, then our rules of thumb come into conflict with one another and they yield no unambiguous result. In the second place, these rules themselves suffer from vagueness, so that in many difficult cases we scarcely know how to apply them. What is it for one concept to be more complex than another? And how are the relative complexities of different statements to be judged, if at all? In the third place, these rules are separate and do not exhibit any clear common principle. They seem to be *ad hoc* rules, rules that might have been formulated by someone who observed the behavior of scientists but who lacked any understanding of the logic according to which scientific reasoning ought to proceed. As disparate rules, they seem to be arbitrary, and there seems to be no clear reason why just these rules should be valid.

It would be philosophically more satisfactory if we could find some common principle that lies at the root of these various disconnected rules of thumb. Had we a common principle, we might hope to lessen the vagueness and ambiguity of these rules and we might be able more clearly to exhibit their reasonableness. We must see whether any such principle can be found.

Nine

Simplicity and Confirmation

IN EARLIER chapters we found reasons for doubting that any version of induction could be satisfactory as the basic principle of nondemonstrative inference. In the last chapter we turned to a different, noninductive sort of principle, the method of hypothesis, according to which a hypothesis is to be regarded as somehow better confirmed the greater the number of observational statements consonant with it that are verified. Besides its general obscurity, this method exhibited two particular defects. In the first place, many an important empirical hypothesis by itself does not imply any empirically verifiable consequents at all (and it may not forbid any observational statement either), and this method does not seem to provide any natural way of confirming these hypotheses. But, in the second place, the method of hypothesis seems to sanction an unnatural way of confirming hypotheses, according to which almost any hypothesis whatever, no matter how absurd, could gain confirmation.

Thus the method of hypothesis as initially formulated is defective. But perhaps its defects are not irremediable; perhaps some plausible modifications in the method might allow it to overcome these difficulties.

In seeking a more satisfactory formulation of the method of hypothesis, let us begin by remembering that there is a significant sense in which the hypotheses that constitute our empirical knowledge fit together to form a system. These hypotheses are not all independent of one another, they are not confirmed one by one in isolation; instead the fate of any one hypothesis is bound up with that of others. In advanced sciences where theories have been erected which are only indirectly connected with observations it is especially clear that a single hypothesis seldom if ever can be confirmed or disconfirmed in isolation from other hypotheses and that much more characteristically it is some rather elaborate set of hypotheses that is put to the test when any new experiment is performed or any new evidence accumulated. In order to obtain experimental confirmation of a hypothesis about the velocity of light, for instance, one must rely upon a whole body of hypotheses about the behavior of one's apparatus and about the general laws of optics. If one obtains an experimental result favorable to the hypothesis which one is particularly trying to test, this is not sufficient conclusively to establish the hypothesis; but this hypothesis and also each of the other hypotheses upon which one has been relying may be expected to gain in credibility thereby. On the other hand, if the result is unfavorable, this does not refute the hypothesis which one was particularly trying to test; it shows only that there is something wrong with the system somewhere, without showing just which particular hypothesis belonging to the system is false. Duhem states this point succinctly when he writes that the scientist "never can submit an isolated hypothesis to experimental test, but only a whole set of hypotheses; when the experiment is in disagreement with his predictions, it shows him that one at least of the hypotheses which constitute this set is unacceptable and ought to be modified; but it does not indicate to him which it is that ought to be changed." [1]

These reflections are commonplace but not unimportant, for this systematic aspect needs to be taken into account if we are to arrive at a suitable view of the logic of confirmation. If we seek a general criterion of confirmation, it will be wiser not to look for a criterion enabling us to determine under what circumstances one isolated hypothesis is better confirmed by the evidence than is another isolated hypothesis. To pose the question in that way would be to neglect this systematic aspect of confirmation. Instead we should seek a criterion that will enable us to decide under what circumstances one set of hypotheses is to be regarded as better confirmed by a given body of evidence than is another set. It will be in virtue of its membership in a confirmed set or system of hypotheses that a particular hypothesis deserves whatever degree of confirmation it possesses; we ought not to direct our attention toward isolated hypotheses in their relation to isolated bits of evidence.

It may be objected that in any concrete case some parts of the available evidence and some among our previously accepted hypotheses will be irrelevant to the question of how well confirmed some particular hypothesis is. For practical purposes this may be so; but so far as theory goes we are not entitled at the very beginning of our inquiries to set aside anything as irrelevant until it has been shown to be so. We cannot be certain ahead of time that any bit of evidence or any already accepted hypothesis may not be involved in the question of how well confirmed some particular hypothesis is.

---

[1] Pierre Duhem, *La Théorie physique: Son objet et sa structure* (Paris, 1906), p. 307.

Induction and Hypothesis

Thus in general it is total systems with which we want to be concerned, total systems embracing all the available evidence and all the hypotheses that may be erected on the basis of that evidence. When we ask whether hypothesis H_1 is better confirmed on the basis of all the evidence than is hypothesis H_2, we ought in effect to be asking whether there is some system of statements which contains H_1 and all the evidence and which is somehow more acceptable than is any system containing H_2 and that same evidence.

What is meant by a system here? The systems with which we are concerned can be thought of merely as consistent sets of empirical statements, each system containing the evidence plus some hypotheses. Since we shall assume that the same logic and mathematics are employed in all the systems of empirical hypotheses with which we deal, we need not bother explicitly to include any purely logical or mathematical statements. Each system will have its characteristic vocabulary of extralogical predicates (finite in number), and all the statements belonging to the system will be formulated by means of these predicates plus the usual logical constants. In addition we shall want to say that any statement which is formulated in the prescribed vocabulary and which is implied by statements that belong to the system is also to belong to the system. In consequence of this, a system will contain innumerably many statements. But we need some way of talking about the system without enumerating its countless members, so we shall want to codify it by selecting some subset of them (a subset that can be effectively characterized) such that by taking these as postulates any statement can be deduced if and only if it does belong to the system. Of course a single system can have various different, but equivalent, sets of postulates.

Now, the first of the two principal defects that we discerned in the method of hypothesis was that many hypotheses do not by themselves imply any verifiable consequents and thus seemingly could not be confirmed in any natural way according to this method. We now can say that the remedy for this defect is to formulate the method of hypothesis so that it shall apply directly to systems of hypotheses and only indirectly to individual hypotheses considered in isolation. We shall then want to say that one hypothesis is better confirmed on the basis of given evidence than is another if and only if there is some system including the former and the evidence which is somehow more acceptable than is any system including the latter and that evidence. If we take this point of view, we shall not have to worry about the relation between isolated hypotheses and isolated bits of evidence in their favor. However, to say that one system can be "somehow more acceptable" than another brings us up against the second of those two principal defects.

II

It will be recalled that the second and more serious defect of the method of hypothesis consisted in the fact that the method seemed to sanction an unnatural procedure according to which any nonempirical hypothesis could receive a relatively high degree of confirmation. The point was that any body of evidence if derivable from a particular hypothesis will also be derivable from each of indefinitely many compound hypotheses logically stronger than that particular one; and the method of hypothesis seems to enjoin us to regard all these hypotheses as equally well confirmed by that evidence.

This difficulty can be rephrased in terms appropriate to the preceding section. Suppose a set of statements constituting a body of verified evidence be given. There are then innumerable different systems to which this evidence can belong, each system containing the evidence plus some array of empirical hypotheses. How are we to choose among these competing systems? It would be absurd to regard all of them as equally acceptable.

Some attempts to cope with this difficulty were noted in the preceding chapter, but none proved successful in general. However, the notion of coherence or logical simplicity seemed to be involved. It would seem that somehow we ought to prefer the system which is simpler, which is more coherent in its structure. If we understood what logical simplicity in some appropriate sense is, we might then be able to contend that, of any two systems to which the evidence belongs, the simpler system is to be regarded as the more acceptable. This way of putting the matter would in effect provide us with a criterion for comparing the degree of confirmation of any pair of hypotheses with respect to a given body of evidence and might suffice to overcome this second principal defect in the method of hypothesis.

In assessing the degree of simplicity of a system of empirical hypotheses, however, we do not want to be concerned with the merely notational aspects of the system. The shapes and arrangements of the inscriptions which occur in its statements are important only insofar as they indicate something about the sort of empirical state of affairs that acceptance of the system would commit us to. Metaphorically speaking, we want to ask, how much does the system require of the world? How complex must the world be in order that the system should hold true? We want to say that the complexity of a system depends upon the complexity which it ascribes to the world. Of two systems each containing the evidence, that system which is content to ascribe the lesser complexity to the world might then be judged the more credible. This is metaphorical, yet it suggests how simplicity and complexity of systems of hypotheses might be relevant to confirmation. But we need to consider what the features of a system are that ought to determine its simplicity.

Simplicity and Confirmation

III

In trying to gauge the simplicity or complexity of a set of hypotheses, it might seem natural to look at the number of its postulates and the structure of each. We might wish to say that one system should rank as more complex than another if it possesses a larger number of postulates, other things equal, or if its postulates are more complicated in structure, other things being equal. This seems natural as a first proposal.

However, this proposal leads to difficulty. It is a trivial fact that one always could replace any given set of postulates by one single postulate which is the conjunction of the given ones; this new single postulate would serve equally well to enable us to deduce any theorem of the system, yet it would be one postulate instead of many and its logical form would differ from the forms of the postulates it replaces. This suggests that the number of postulates a system has, or the form of these hypotheses, will not provide any easy clues to the degree of simplicity of the system. Indeed, any single system can be codified in many different though equivalent ways: there are innumerable different sets of postulates which might be selected for it. And these different sets of postulates will differ in number of postulates, and they may differ with regard to the logical forms of their respective postulates. Since we wish to elicit a sense of complexity and simplicity which will be relevant to the credibility of a system of hypotheses, we want to be able to regard a single system as having a definite degree of complexity, however the system may happen to have been codified. Thus, since a system can be codified in any number of different ways with different numbers of postulates having different forms, it does not seem helpful to look to the number and structure of the postulates when we seek symptoms of simplicity. The number of postulates a system has and their logical structure would

Induction and Hypothesis

seem to be unilluminating as indications of the simplicity or complexity of the system itself, for these factors do not depend just upon the nature of the system but rather upon the specific manner of codification that happens to have been chosen.

If it will not do to count postulates, another suggestion which seems natural is that in order to gauge its simplicity perhaps we should count the predicates belonging to the basic extralogical vocabulary of the system; perhaps we should say that one system is simpler than another just in case it has fewer primitive predicates. Here there will arise no question of alternative formulations of the same system: no matter how it is codified, a system must always retain its characteristic vocabulary—for if this basis is altered a new and different system results. And certainly it does seem rather plausible to suppose that the number of primitive predicates (or we might say, the number of undefined concepts) employed in a system is a clue to its complexity. We do incline to think of a system containing fewer primitives as simpler than one which contains a larger number, for when we seek conceptual economy in theories, fewness of primitives surely is in question.

However, if we were to follow this suggestion and regard the mere number of primitives of a system as the key to its complexity, we should again be in trouble. This is because transformations of a trivial sort enable us to compound together all the primitives of any basic vocabulary so as to obtain a single primitive predicate which is capable of doing exactly the same job that the various predicates of the original basis do. This procedure involves replacing the original system by another different one, of course; but the new system differs from the old one only in trivial notational respects, so that we certainly should not wish to regard the two systems as differing in simplicity—not if we wish to regard simplicity as a mark of credibility.

172

Simplicity and Confirmation

As an example of how this trivial transformation might be carried through, let us consider a system S_1 whose vocabulary consists of three one-placed predicates 'F,' 'G,' and 'H.' We need not consider what these predicates mean or what statements containing these predicates the system includes; let us assume only that the system does include statements to the effect that none of these three predicates is true of nothing. Thus in order to describe certain facts this system uses these three predicates; but it is easy to construct a new system S_2 which will describe the same facts yet which will need only one predicate in place of the three of the previous system. We could let 'Kxyz' be this predicate; it would be so explained as to be true of a sequence of three things x, y, and z if and only if x is an F, y a G, and z an H. By using this new three-placed predicate 'K,' the statements belong to the earlier system all can be reformulated in accord with these rules:

For 'Fx' read '$(\exists y)(\exists z)Kxyz$';
for 'Gx' read '$(\exists y)(\exists z)Kyxz$';
for 'Hx' read '$(\exists y)(\exists z)Kyzx$.'

The new system S_2 is to contain all and only those statements which can be got from statements of S_1 by means of these rules. If we regarded the simplicity of a system as determined by the mere number of its primitive predicates, then we should be obliged to rate S_2 as simpler than S_1; but obviously this would be implausible, for these two systems differ only in a trivial way. S_2 has been obtained from S_1 by a procedure which would always work and in effect S_1 and S_2 say just the same thing. Therefore, we should not wish to regard S_2 as simpler than S_1, and we must reject the notion that mere number of predicates is a reliable index of complexity.

Induction and Hypothesis

IV

Instead of trying to estimate the acceptability of a system merely in terms of the number of primitives it has, a more sophisticated approach is to consider the logical structure which these primitive predicates possess. Goodman has developed a theory of the logical simplicity of predicates, a theory in terms of which the relative simplicity of sets of predicates can be compared; and he has advocated its use as one criterion to be employed in choosing among competing systems.[2] Can this theory be used to guide us in choosing among sets of empirical hypotheses?

Roughly speaking, Goodman's scheme is as follows. In considering the simplicity of a set of predicates it is not the particular predicates with their special meanings that matter; what matters is that they are a set of predicates having formal characteristics of a specific kind. Reflecting upon the trivial mechanical method of replacement (as witnessed in the preceding section), Goodman enunciates a "principle of replaceability" to the effect that if every set of predicates of one kind is replaceable by some set of predicates of another kind then the first kind cannot be more complex than the second. This principle he conjoins with various further principles about the effect which symmetry and self-completeness have upon the simplicity of sets of predicates. His theory then is strong enough to permit general comparisons between systems of all kinds.

It is a consequence of this formulation of the theory, however, that interrelations among various predicates of a basis cannot be relevant to the measurement of the complexity of the system to which they belong. That is, if the postulates of a system guarantee that two of its primitive predicates must have

---

[2] Nelson Goodman, "Axiomatic Measurement of Simplicity," *Journal of Philosophy*, III (1955), 709–722.

the same extension, or that the extension of one must include that of another, these facts in no way affect the complexity of that system. Now, a theory of simplicity which leaves out of account all interrelations among different predicates and takes account only of the number of places the predicates have, of their symmetry and of their self-completeness, may be a theory having fruitful applications to some problems. But it is not a theory which can be applied to our problem of choosing among competing systems of empirical hypotheses.

This is because we do want to be able to say that a generalization, for instance, can sometimes be confirmed by empirical evidence; and—if we are relying upon a notion of simplicity as our clue to confirmation—we shall want to say that a system which contains this empirical evidence can be simplified if the generalization is added to that system. That is, we need a notion of simplicity according to which a system containing a generalization such as 'All F are G' (asserting in effect that the extension of 'G' includes the extension of 'F') can be simpler than an otherwise similar system which does not contain the generalization. Generalizations and other familiar sorts of empirical hypotheses do assert relationships between predicates, and we want to say that adoption of these hypotheses sometimes can simplify our systems. But Goodman's latest formulation of his theory of simplicity treats these relationships among predicates as indifferent to simplicity. To be sure, in earlier formulations of his theory Goodman used a stronger form of the principle of replaceability and did allow some such relationships to count;[3] but in order to prevent anomalies it is necessary then to stipulate that the principle of replaceability shall apply only to disjoint kinds of bases, and a consequence of this is that a

---

[3] Goodman, *The Structure of Appearance* (Cambridge, Mass., 1951), ch. iii. See also Lars Svenonius, "Definability and Simplicity," *Journal of Symbolic Logic*, XX (1955), 235–250.

generalization is not superior to its denial at the job of simplifying a system into which it may be introduced. Thus it seems that Goodman's notion of simplicity cannot be used to provide us a criterion of confirmation.

V

We saw that for gauging degrees of confirmation it seemed necessary to consider whole systems of hypotheses, not just single isolated ones. And we saw that in choosing among competing systems of hypotheses it seemed proper to appeal to some sort of logical simplicity. But what notion of simplicity is appropriate? The detailed theory of simplicity which Goodman has developed seems not to be relevant here, and we need to look for simplicity of a somewhat different sort. Some work of Kemeny's may be helpful in this connection, for he has suggested as one possible notion of logical simplicity an idea related to his logical measures.[4] Basically his idea is that the logical measure of a system indicates its richness or complexity, and hence that the less its logical measure is, the greater is its simplicity. The scheme can be informally explained as follows.

Let us suppose by way of example that we have a language in which the only extralogical terms are the predicates 'is a crow' and 'is black.' Now suppose, for instance, that the universe contains just two individuals *a* and *b*. If this is all we know about the situation, then we cannot say whether either *a* or *b* actually is a crow or is black (in this language we do not even have names for these individuals); but we do know that (with respect to being a crow and being black) exactly sixteen constitutions of this universe are possible—in Kemeny's terms, a two-membered universe provides sixteen models for this language. These are:

---

[4] John Kemeny, "Two Measures of Complexity," *Journal of Philosophy*, LII (1955), 722–733.

Simplicity and Confirmation

1. *a* and *b* both are black crows.
2. *a* is a nonblack crow, *b* is a black crow.
3. *a* is a black noncrow, *b* is a black crow.
4. *a* is a nonblack noncrow, *b* is a black crow.
5. *a* is a black crow, *b* is a nonblack crow.
6. *a* is a nonblack crow, *b* is a nonblack crow.
7. *a* is a black noncrow, *b* is a nonblack crow.
8. *a* is a nonblack noncrow, *b* is a nonblack crow.
9. *a* is a black crow, *b* is a black noncrow.
10. *a* is a nonblack crow, *b* is a black noncrow.
11. *a* is a black noncrow, *b* is a black noncrow.
12. *a* is a nonblack noncrow, *b* is a black noncrow.
13. *a* is a black crow, *b* is a nonblack noncrow.
14. *a* is a nonblack crow, *b* is a nonblack noncrow.
15. *a* is a black noncrow, *b* is a nonblack noncrow.
16. *a* is a nonblack noncrow, *b* is a nonblack noncrow.

We do not know which is the true state of affairs, but it must be one among these listed. Let us now consider two statements which can be expressed in this simple language: 'Anything that is a crow is black' (according to this statement, the extension of 'is black' includes that of 'is a crow') and 'Anything is a crow if and only if it is black' (according to this statement the two predicates have the same extension). The statement 'Anything that is a crow is black' comes out true in cases 1, 3, 4, 9, 11, 12, 13, 15, and 16, because in each of these cases there is no crow that is not black. The statement 'Anything is a crow if and only if it is black' comes out true only in cases 1, 4, 13, and 16, for in every other case there is something which is black without being a crow or which is a crow without being black. Thus, for two-membered universes, the former statement can come out true in more ways than can the latter statement. And in this little language a system having the former statement as its sole postulate would have more models in any two-membered universe than would a system having the latter statement as its sole postulate. Of course, any system containing the latter statement as a postulate would necessarily contain the former statement as a theorem; but not conversely.

We have been considering a very simple language; but this same sort of procedure can also be applied to languages which contain more predicates and predicates with more places. In general when we ask how many ways a statement (or set of statements) in a given language has of coming out true provided the universe is n-membered, we are asking the following question: how many different ways are there in which an n-membered universe can be constituted so as to provide an extension (not necessarily non-null) for each predicate of the language—these extensions being chosen in such a way that the statement (or all the statements of the set) are true. A one-place predicate must have as its extension some class of individual things selected from the universe (the individuals the predicate is true of); a two-placed predicate must have as its extension a class of ordered pairs of individuals (the pairs that this relational predicate holds true of); and a k-placed predicate must have as its extension a class of k-ads of individuals. No matter how many predicates the language contains (so long as the number is finite) and no matter how many places each of these predicates has (so long as each has only a finite number of places), there is always a definite answer to the question, in how many ways could such-and-such statements of the language come out true if the universe contains just n numbers?

As we saw earlier, Kemeny wants to say that the "logical measure" of one statement (or set of statements) is greater than that of another if the one can come out true in more ways than can the other. However, we have to allow for universes of various sizes; we do not want to restrict ourselves to universes of some specific size. Kemeny adopts the following stratagem:

let us say that the logical measure of one statement (or set of statements) is greater than that of another just in case, for every sufficiently large n, the former statement can come out true in more ways than can the latter in an n-membered universe.

With regard to the two statements that we were considering, we saw that, in that simple language, 'Anything that is a crow is black' comes out true in more ways in a two-membered universe than does 'Anything is a crow if and only if it is black.' But now we can go on to observe that, whatever the size of the universe may be (so long as it is sufficiently large—in this case, having more than zero members—and so long as it is not infinite), the former statement always will come out true in more ways than will the latter statement in universes of that size. The relationship is evident since, if the universe is of size n, the former statement can come out true in 3^n ways while the latter can come out true only in 2^n ways. Let us think again of two systems, one having the former statement as its sole postulate and the other having the latter statement as its sole postulate. This first system has two one-placed predicates so related that the extension of one includes that of the other; the second system also has two one-placed predicates, but it requires both to have the same extension. Here we shall want to say that the first system allows greater leeway, more variety and richness to its models; whereas the second system imposes a greater uniformity upon its models—and its two predicates, because they are coextensive, in effect do the job of just a single one-placed predicate.

At this point is it worth while to notice an interesting objection of Goodman's that might be levied against any attempt to use this notion of logical measures as a criterion of simplicity.[5]

---

[5] Professor Goodman kindly communicated this objection to me.

The objection is that two ordinary one-placed predicates ought always to be of equal simplicity; to add one of them to a system ought to increase the complexity of the system no less and no more than would adding the other. But it does not always work out this way with Kemeny's logical measures. For instance, consider the predicates 'poorer than at least 90 per cent of the population' and 'poorer than at least 50 per cent of the population.' These are two ordinary one-placed predicates which seemingly ought to count as equally simple; yet different logical measures will be associated with them, for the latter will allow far more models than will the former in universes of sufficiently large size. Does this show that logical measures cannot provide a suitable criterion of simplicity? From Kemeny's point of view, it would seem that one ought not to accept this objection; instead one ought to insist that whenever we possess any advance information about one of our predicates we ought explicitly to state this information in nonempirical "meaning postulates" which we add to our system. Thus if we add to your system the predicate 'poorer than 50 per cent of the population,' then we should add also a meaning postulate expressing the fact that the extension of this predicate must be less than half the universe; and surely it is not too surprising that this piece of information should affect the complexity of the system. If this reply to the objection be tenable, perhaps we may proceed with our supposition that logical measures can serve as a criterion of simplicity.

If we wish to use this notion of simplicity as a guide for choosing among competing systems of empirical hypotheses, then we shall want to imagine ourselves to possess a definite language adequate for expressing any empirical statement with which we shall be concerned. This language will contain the logic of truth-functions and quantification plus a finite number of extralogical predicates (no one can expect to understand more than a finite number of undefined predicates). We shall have a number of verified observational statements, and these, since we know them to be true, must be incorporated into any system which we could consider adopting. But there will be innumerable different systems of hypotheses that do contain this observational evidence. What we shall want to do is to use this notion of simplicity in order to determine the relative degrees of acceptability of these competing systems. In comparing any two systems each of which does contain the verified evidence, we shall want to say that the system with the lesser logical measure is to count as the simpler and hence as the more acceptable. This does give us a definite criterion in terms of which to decide about the acceptability of systems of hypotheses. And if we wish to compare the relative degrees of confirmation of two individual hypotheses with respect to a body of evidence, we may say that one of these hypotheses is better confirmed than is the other if and only if the former hypothesis plus the evidence belongs to some system which is simpler than is any system to which the latter hypothesis and the evidence belong.

Inevitably it will be asked why we should prefer simpler systems. Why should we suppose that a system which is simpler in this sense is more acceptable than one which is complex? Why should it be more reasonable to expect a simple system to be true than to expect a complex one to be true? This question perhaps is too fundamental to admit of any sharp answer. Certainly we cannot prove that preference for simpler systems must lead one to true beliefs, even "in the long run." Nor does it seem to be possible to offer any convincing argument in its favor based upon the statistical syllogism, upon the principle of indifference, or upon any simple rule of induction. However, we can say this: if one system is simpler than another, in this sense, then the simpler one "says more," it has "more content," because it excludes a greater number of possible models; therefore it runs more risk of being contradicted by the evidence. A system which takes a risk yet survives deserves more credit, it earns more credibility, than does a system which survives but says less and thus has taken less risk. This reflection may help make preference for the simpler system seem intuitively more plausible. But even if no such justification is satisfactory, that might not necessarily be a drawback, for if a notion of simplicity like this really is to be regarded as fundamental to nondemonstrative reasoning, then it would not be surprising if it were so fundamental as not to admit of being justified in any clear way by appeal to any other principle of probability or induction. This is not necessarily a defect, for one need not expect to justify every principle in terms of some other.

VI

If we are going to think of systems of empirical hypotheses as more credible the simpler they are, then we need to have a notion of simplicity such that when hypotheses of certain appropriate sorts are added to a system the system thereby becomes relatively simpler than would otherwise be the case, while if hypotheses of inappropriate kinds are added the system does not become relatively simpler. Does the notion of simplicity suggested in the preceding section accord with this?

Ordinary generalizations, statements of the form 'All F are G,' constitute one important kind of hypothesis which we should like to be able to confirm. Let us compare the effect of adding a generalization to a system with the effect of adding its denial. Suppose we have a language whose vocabulary includes the one-placed predicates 'F' and 'G,' and suppose that we are wondering what hypothesis we ought to adopt about whether F's are G's. Assuming that no F's have been observed which are not G's, ought we to adopt the hypothesis that all F's are G's or should we prefer the hypothesis that there are F's which are not G's? That is, how should we choose between a system on the one hand which contains 'All F are G' as its only hypothesis about F's and G's and a system on the other hand which contains 'Some F are not G' as its only hypothesis about F's and G's? The answer is clear, if we embrace the notion of simplicity of the preceding section as our criterion for choice. If we look into the algebra of the matter, we find that in a universe of any given size the former system will have fewer models than the latter system has. Moreover, if we have evidence about F's and G's, then this evidence of course must be incorporated into each system; and the greater the number of favorable instances of the generalization (that is, the more cases we observe of things that are not F without being G; and in order to be recognized as different instances, these cases must be observed also to differ from one another), the relatively greater will be the disparity between the number of models which the former system has in a universe of given size and the larger number of models which the latter system will have in a universe of that size. Thus, in a case like this, the generalization does better than its negation at the job of reducing the number of models of the system to which it belongs. And the more favorable instances there are, the greater is the superiority of the generalization to its negation at this job. Thus this notion of simplicity enables us to say that the generalization is relatively better confirmed than its negation, and relatively more so the greater the number of favorable instances.

Another important type of hypotheses which we should like to be able to confirm consists of relational hypotheses, that is, multiply quantified hypotheses in which occur predicates having more than one place. Does the proposed notion of simplicity enable us to maintain that the introducing of a hypothesis of this type into a system may sometimes be justified because of the relative increase in simplicity? There are many different forms which such hypotheses may have, but as an elementary example let us consider a hypothesis of the form '$(x)(Fx \supset (\exists y)(Gy.Rxy)$'; a hypothesis of this form asserts that every F bears R to some G or other. Suppose we have evidence that some F's do bear R to G's. Any system that we adopt must contain this evidence, but should we add some hypothesis as well? The hypothesis that no F bears R to any G has been falsified by our evidence, so we cannot adopt it; but we may choose between the hypothesis that every F does bear R to some G or other and the competing hypothesis that some F's do not bear R to G's. Again the algebra becomes complicated, but if we look into the matter, we shall find that for a universe of any given size (except the smallest ones) the evidence conjoined with either of these two hypotheses would have fewer models than would the evidence alone; moreover, the former of these two competing hypotheses will yield a system having fewer models than that yielded by the latter hypothesis conjoined with the evidence. And thus, under these circumstances, we should regard the former as the simpler and hence as the preferable hypothesis. And the observation of additional F's that do bear R to G's will strengthen our evidence in such a way as to increase the disparity in favor of the former hypothesis. This indicates that the proposed criterion of confirmation may allow for the confirming of relational hypotheses such as this.

Furthermore, the fact that a relational hypothesis like this one can be confirmed shows that it may be possible, according to this scheme, for hypotheses implying the existence of unobserved entities to be confirmed also. An elementary case of this kind might arise in the following way. Suppose we have observed that two things x and w are both F's, but that x bears R to something y that is a G, while w does not bear R to this y. Let us suppose that, as proof of the distinctness of these three things, we have observed that y bears R to itself but that neither x nor w bears R to itself. Under these circumstances should we adopt a hypothesis? Just as before, we may consider the two competing hypotheses that every F bears R to some G and that some F's do not bear R to any G. Either of these hypotheses when combined with the evidence will yield a simpler system than would be constituted by the evidence alone; but as before, the former hypothesis yields a system simpler than that obtained by adding the latter hypothesis to the evidence. And here, if we adopt the hypothesis that every F bears R to some G, we are thereby adopting also the implied hypothesis that there exists something z which is a G and to which w bears R. Our evidence shows that this z must be distinct from x, y, and w. Thus we have obtained some confirmation of a hypothesis which implies the existence of an unobserved entity.

One further important kind of hypothesis which we want to be able to confirm consists of hypotheses about mathematical relationships between quantitative properties; these are important in cases in which observations indicate a number of points on a graph and we wish to establish a hypothesis about the curve that should be fitted through these points. Examples like this involve much more complicated languages than do the elementary examples we have so far been considering, for in order to state a hypothesis of this sort one needs to be able to speak of numbers and of functions of them; our logic would have to be supplemented so as to yield the required mathematical locutions. But although examples are bound to be rather elaborate when worked out in detail, we should like to have some advance indication whether this criterion of confirmation may prove appropriate to them. That it may is perhaps suggested by the following very informal consideration.

Suppose that we have two quantitative properties, and suppose that we have established that, say, a value of 1 for the first is associated with a value of 2 for the second, of 2 for the first with 3 for the second, of 3 for the first with 4 for the second, and so on. Now we wonder what kind of curve to plot as a generalization about the relationship between these two quantitative properties. We might conjecture that the relation is a linear one; or we might suppose it to be some more complicated relationship which mathematical ingenuity may invent. How do these competing hypotheses compare with respect to their simplicity? This much at least we can say: the hypothesis that this relationship is linear is a hypothesis which, when conjoined with the evidence, is consonant with only one type of structure of the universe, with respect to these two quantitative properties; whereas, the hypothesis that the relationship should be represented by an equation of some more elaborate form is a hypothesis which, even when conjoined with the evidence, still is consonant with many different types of structure of the universe. The former hypothesis may be expected to exclude more models than would the latter, and in this sense the former should be accounted the simpler and more acceptable hypothesis. This informal consideration at least does suggest that the proposed criterion of confirmation may give us a reason for preferring the hypothesis here which intuitively and mathematically does seem the simpler.

VII

We have considered several of the principal kinds of hypotheses that we want to be able to confirm, and it seemed that the proposed criterion was able to account for the confirming of hypotheses of these kinds. However, there is an objection which may be raised; it may be contended that this criterion of confirmation is obliged to give a very high degree of confirmation to a certain kind of hypothesis which really ought not to be confirmed. The kind of hypotheses at stake here comprises those which assert some upper limit to the number of individuals in the universe. Suppose, for instance, that we have observed a certain number of things, and we wonder what hypotheses to base upon this observational evidence. It would appear that there always will be one outstandingly simple hypothesis: the hypothesis that there do not exist more than this certain number of things in the universe. If our evidence guarantees that there are at least k things in the universe, then why not adopt the hypothesis that there are at most k things? This hypothesis will simplify our system to such an extent that our system will have zero models in any sufficiently large universe (larger than k); and therefore it would seem that there could be no simpler hypothesis. Yet of course it is unreasonable to claim that this is the best confirmed of all hypotheses.

This objection can perhaps be met in the following way. In order to formulate the hypothesis that there are at most k things, one would have to employ the sign of identity. One would assert that if there are more than k things then some of them are identical with one another. But is a hypothesis of this character a proper empirical hypothesis? Is the bare assertion that two things are identical an assertion that has empirical significance? Surely we should have to maintain that, so far at least as empirical discourse goes, to assert of two things that they are identical must always be to assert that they have certain characteristics in common: they occupy the same space, or they are stages of a single spatiotemporal chain, or something of that sort. The bare assertion that two things are identical cannot be empirically significant unless it means something to this effect. Whatever may be the case *in rerum natura,* at least in our hypotheses there can be no distinction without a difference and no identity without common characteristics. But if this be granted, then we cannot countenance the hypothesis that the universe contains at most *k* things; this hypothesis involves bare statements about identity, statements which have no empirical significance and therefore which cannot properly be introduced into any system of empirical hypotheses. One is entitled to frame all the hypotheses one likes about all things of such-and-such kinds or about some things of such-and-such kinds—but these hypotheses will not turn out to have no models in sufficiently large universes.

After all, one never could observe that two things merely are identical or merely are distinct. What one observes is that there are things which differ in certain respects or which are alike in certain respects, and only thence does one infer that there are distinct or identical things. Therefore, it appears not unreasonable to set aside this seemingly serious objection.

VIII

Let us turn briefly to another quite different but rather challenging objection that Goodman has raised, an objection which aims to show that no criterion of confirmation like that just discussed, or indeed like any discussed in preceding chapters, can be adequate.[6] Here are involved puzzles like that of the grue and bleen emeralds.

The puzzle may be stated in this way: suppose that we have observed a large number of emeralds of a variety of sorts and all of them have been green. It would seem natural, on the basis of this evidence, to accept as probable the generalization that all emeralds are green. However, let us define 'grue' to mean 'green prior to the year 2000 and blue thereafter.' Now consider the hypothesis that all emeralds are grue; why should we not accept this hypothesis rather than the more prosaic one that

---

[6] Goodman, *Fact, Fiction, and Forecast* (Cambridge, Mass., 1955), ch. iii.

all emeralds are green? All the numerous emeralds that we have observed have been grue, so we seem to have an equal number of observed cases in favor of each generalization. Clearly we should not want to accept both hypotheses, for they yield conflicting predictions about emeralds after the year 2000. But what reason have we for preferring one generalization to the other?

Some philosophers, Carnap for instance,[7] have tried to escape from Goodman's trap by stipulating that only predicates which are purely qualitative may figure in inductive generalizations. 'Green' is said to be a purely qualitative predicate, whereas 'grue' because it involves reference to a particular date is not purely qualitative; so we accept the hypothesis that all emeralds are green and reject the hypothesis that all emeralds are grue. At first blush this may sound promising, but Goodman's reply is that we do not have any clear way of distinguishing between predicates that are purely qualitative and those that are not. If someone were to present you with a list of predicates, how would you go about determining which are purely qualitative?

The answer that I should like to give is that there really is this difference between a predicate like 'green' and one like 'grue': at least in some cases, one can verify by direct observation that a thing is green; whereas one never could conclusively verify that a thing is grue. It is always an hypothesis that a thing is grue, because grueness depends upon the date; perhaps by looking at the calendar you can verify that the calendar reads so-and-so, but you cannot verify that the calendar is right. The point is that to tell whether a thing is green you need only determine its color; while to tell whether a thing is grue you need to determine the color of the thing and also whether the date is before or after the year 2000.

---

[7] Rudolf Carnap, "On the Application of Inductive Logic," *Philosophy and Phenomenological Research*, VIII (1947–1948), 133–148.

Induction and Hypothesis

Goodman has an answer to this. His answer is that there is a thorough symmetry between 'green' and 'grue.' In terms of 'green' and 'blue,' we can define the predicates 'grue' and 'bleen': 'grue' means 'green prior to 2000 and blue thereafter' and 'bleen' means 'blue prior to 2000 and green thereafter.' However, suppose there were somebody who spoke the grue-bleen language instead of the language of green and blue; he could define 'blue' to mean 'bleen prior to 2000 and grue thereafter' and 'green' to mean 'grue prior to 2000 and bleen thereafter.' He would claim that in order to tell whether something is grue you need only determine its color; whereas to determine whether a thing is green you need to determine its color and the date.

Here one is tempted just to pound the table and insist that green a single real color and grue is not. This is what makes the puzzle interesting—it leads us to reflect about the relation, if any, between language and reality. Goodman's view is that we prefer the hypothesis 'All emeralds are green' to the hypothesis 'All emeralds are grue' just because the predicate 'green' happens to be entrenched in our language; if our grandfathers had spoken the grue-bleen language instead, then we should prefer the hypothesis 'All emeralds are grue.' But if we wish to adhere to the general point of view expressed in earlier chapters, then we shall not be satisfied with this and we shall want to insist that the formulation of our observational statements is not a matter of such arbitrary convention. We shall want to try to maintain the claim that there is more of a connection between language and reality than Goodman's view seems to allow. If we do maintain the claim that 'green' is an observational predicate but 'grue' is not (because one can verify by direct inspection that a thing is green, but cannot verify by direct inspection that it is grue), then we can explain our choice of the hypothesis 'All emeralds are green' in the following way.

Simplicity and Confirmation

We can argue that since we never verify for certain that anything is grue there is no reason for introducing 'grue' into our language at all, but that if the predicate is introduced then the simplest and preferable system of hypotheses will be one which includes a hypothesis that will serve to prevent 'grue' from adding to the number of models that the system has—this might, for instance, be a hypothesis to the effect that 'grue' is true of nothing. By conjecturing that 'grue' is true of nothing, we may obtain a system having the minimum logical measure. But under these circumstances the hypothesis that emeralds are grue will become trivial and uninteresting as compared with the hypothesis that emeralds are green; for since greenness is a property which we observe that some things possess and some do not, the hypothesis that all emeralds are green is capable of effecting an interesting and nontrivial simplification in the system.

Ten

Concluding Remarks

I

A CRITERION of confirmation which is a development of the method of hypothesis and which has no basic affinity with induction has been tentatively proposed in the preceding chapter. We saw that according to this criterion it seemed to be possible for hypotheses of various important kinds to gain confirmation—at any rate, when combined with favorable evidence these hypotheses seemed to contribute more to simplicity than could their negations. Of course this is merely an indication, and not at all a proof, of the adequacy of the proposed criterion.

Some philosophers, it is true, would assert that an ideal criterion ought to determine a numerical value for the degree of confirmation of any hypothesis with respect to any possible body of evidence. This proposed criterion, however, is merely a comparative criterion, and is only a partial one at that. It enables us to compare the degrees of confirmation of two hypotheses with respect to a single body of evidence; but it does not enable us to assign numerical values to degrees of confirmation, nor does it even help us to compare the degree of confirmation of hypotheses with respect to different bodies of evidence. But one might plead, in extenuation of this, that the proposed criterion does aim to do the one thing which is most important to us; it does help us to compare competing hypotheses. One body of evidence is all that we ever have—so perhaps the fact that this criterion does not tell us how to compare degrees of confirmation with respect to various bodies of evidence is not too serious a defect. And neither in everyday life nor in scientific practice is it customary for the degrees of confirmation of hypotheses to be so clear-cut as to admit of being assigned numerical values; so perhaps it really is too much to ask that a criterion should determine such numerical values.

It may be felt, however, that the proposed criterion is arbitrary, that it has been artificially constructed, and that there really is no reason for accepting it as our guide in comparing the rational credibility of sets of hypotheses. Certainly there does not seem to be any empirical evidence in favor of this criterion as opposed to other possible ones; there seems to be no good reason for considering it to be analytic, in any useful sense of that unfortunate term; thus it would appear that the only thing this criterion could have in its favor might be some claim to self-evidence. Yet it does not really seem that the criterion is self-evident, for no necessity leaps to the eye in connection with it.

To this it must be replied that even when a philosophical doctrine actually is correct it may often be that its correctness is not apparent to the casual eye. What is required, surely, is that the whole subject be taken under consideration, all the interconnected problems being considered together, and that there be elicited a doctrine which on the whole will yield the most plausible account of all the matters involved. In philosophy, as in empirical inquiries, our problem always is to choose among competing theories; and one theory gains credibility only as its competitors are shown to be relatively less plausible. If we cannot find a doctrine that is intrinsically plausible, then at least we can seek the least implausible doctrine. Naturally the view chosen, which obtains its credibility from its contrast with the incredibility of competing views, may seem arbitrary and implausible when isolated and considered in abstraction; it ought rather to be viewed in its contrast with competing views over the whole range of philosophical issues to which it pertains.

What was argued in earlier chapters was that there must be some logical principle underlying the confirmation of empirical hypotheses; were confirmation a matter of caprice, stultifying skepticism about empirical knowledge ought necessarily to ensue. What has now been argued is that the proposed criterion may perhaps be more adequate as a formulation of this fundamental principle than is any other criterion which seems to be available. This criterion is suggested as fundamental to nondemonstrative inference, a criterion having a vital role to play in the edifice of empirical knowledge. It is a criterion claiming to be able to do the work that each of the theories of induction and confirmation previously examined did not seem able satisfactorily to do: it claims to make it possible for evidence of the available kind to confirm hypotheses of the sorts that science and common-sense knowledge require.

Naturally it would be absurd to pretend that this claim has been established; the preceding chapters have done no more at best than to delineate the proposed criterion and its main competitors and to suggest some of the apparent virtues and defects of each. The proposed criterion of confirmation may indeed be defective; but if it is so, it ought to be rejected on account of its technical inferiority to some better criterion, not on account of a mere want of immediate self-evidence.

II

If the doctrine of confirmation which has been suggested be accepted, some conclusions of more general philosophical interest emerge. These relate to the nature of explanation, to the kind of terms that should appear in hypotheses, and to the metaphysical status of the entities referred to by these hypotheses.

In the first place, we shall conclude that to provide an acceptable new scientific explanation must be to effect some simplification in one's system of hypotheses. That is to say, a new statement not belonging to the system which we have based upon our evidence ought to be introduced into our system only if by so doing we bring about some relative increase in the simplicity of that system. To introduce a statement which does not effect any such simplification would be to adopt a hypothesis which is not supported by the evidence and which could not yield a legitimate scientific explanation. If we adopt this view, then we must reject as misleading an example which Braithwaite sets store by: he supposes that if the evidence has led us to construct a scientific theory in which three predicates occur then it would be legitimate to provide an explanation of this theory by deducing it from a new theory in which occur these same three predicates plus three new ones.[1] But such a procedure is going to be unsatisfactory in terms of our criterion, for the new theory is far more complex than the old in virtue of the fact that added predicates greatly increase the number of models the system will have in universes of any given size. Indeed, Braithwaite's procedure, if pursued further, would lead

---

[1] R. B. Braithwaite, *Scientific Explanation* (Cambridge, 1953), pp. 63 f.

to fantastic results; for if we make the step Braithwaite suggests, we might just as well then go on to "explain" our six-predicate theory by means of a twelve-predicate theory, and that in turn by means of a twenty-four-predicate theory, and so *in indefinitum.* Such "explanations" are not of any genuine scientific value, because they are of a sort which could always trivially be provided; they effect no simplification at all.

A second, related conclusion which we may draw from the doctrine of confirmation which has been proposed concerns the nature of the terms which are to appear in legitimate scientific hypotheses. Some writers have argued that predicates may occur in scientific hypotheses which have not been observed to have any instances; they cite ostensible examples of such theoretical terms from physics, psychology, and the like. But we shall be obliged to regard such claims with suspicion, for they seem to embody a dubious conception. It will be recalled that a system of scientific hypotheses based upon a given corpus of evidence must include, in its basis, all the predicates occurring in that evidence. But all those predicates will necessarily be predicates observed to have instances, for an inapplicable primitive predicate cannot occur (unless vacuously) in the statement of the evidence. We seek to erect upon this evidence the simplest possible system, however; could the introduction of a new primitive predicate, a predicate not occurring in the evidence, ever contribute to simplicity? In answering this question, we may distinguish two cases. Suppose the new predicate were assumed by hypothesis to be true of nothing, or true of everything, or coextensive with some other predicate already occurring in the system. In this case the new predicate will not increase the complexity of the system, for it will not generate any added models. However, a new predicate which is assumed to be inapplicable or to be universal or to be coextensive with some other predicate is a predicate which really adds nothing useful, which enables us to divide up things in the world in no new way, and it would be pointless for us to introduce such a predicate when the evidence does not require us to do so. Suppose, on the other hand, that the new predicate were not assumed to be universal or inapplicable or coextensive with another. In this case, the additional predicate would increase the number of models the system has in an n-membered universe by a factor of 2^n, since the new predicate could be either true or false of each of the n individuals. But such a new and unnecessary predicate which adds to the complexity of the system ought not to be introduced, for it weakens the degree of confirmation of the hypotheses belonging to the system. Thus, according to the view being proposed, in neither case should any predicate not occurring in the evidence be introduced into any scientific hypothesis that we are going to adopt (though of course this applies only to undefined predicates; there is no objection to predicates definable by means of predicates occurring in the evidence).

This conclusion accords with the point made in connection with formalism: any system containing theoretical predicates can be replaced by some system or other in which occur no such predicates, the new system being such as to include all and only observational statements which are included in the old system. This fact shows that the introduction of theoretical predicates is a procedure which cannot be of any genuine logical significance, for it cannot increase the observable consequents of the theory and it is a procedure which is always trivially possible.

We have argued that no term ought officially to occur in an acceptable hypothesis (defined terms occur unofficially, so to speak) which does not occur in the evidence and that therefore it is misleading to speak of theoretical terms as though they deserved to be imported *ad libitum* into science. It may be objected that in scientific theories there do occur many terms which nowhere occur in the evidence. Perhaps this is correct.

However, what is being suggested is that any extralogical term legitimately being used in science or in everyday empirical knowledge in principle ought somehow to admit of being defined by means of observational predicates which do occur in the evidence. Such a definition may be quite complex and difficult to supply, but if it cannot be provided, then we are entitled to doubt that the term in question is being used legitimately. However, it would be a formidable task indeed to defend this point of view by carrying out an examination of the manner in which apparently theoretical terms actually do or ought to function in science, for instance in quantum mechanics.

Lest there be misunderstanding, it should be noted that this is not reductionism which is being proposed. Reductionism was the view that any significant empirical statement must admit of being translated into a statement none of whose terms purports to denote any but directly observed things. What is here being proposed is indeed that predicates applicable to observed things should be the foundation of all discourse about matters of fact; but this is not the same as to say that all discourse must be discourse about observed things—for the properties and relations which are possessed by directly observed things may be possessed also by entities which are not directly observed. The view being proposed is less stringent than reductionism, and it claims merely that whatever individual things there may be, observed and unobserved, there is no reason for attributing to them any extralogical properties or relations other than those which are exhibited by directly observed things (or which are definable in terms of properties and relations they exhibit).

In effect, the view being proposed is merely empiricism. In place of "nothing is in the mind which was not first in the senses," we have been arguing that no predicate ought to occur officially in any accepted hypothesis which was not first in the observational evidence. Though based on a doctrine of confirmation rather than upon a doctrine of meaningfulness, the result is much the same, for metaphysical hypotheses are eliminated. Consider such hypotheses as 'If there are entelechies, then they cause growth,' 'If there is luminiferous ether, then it transmits electromagnetic vibrations,' or 'If there is a World Soul, then it is objectified in History.' These hypotheses contain terms which do not occur in observational evidence and which can scarcely be regarded as definable by means of observational predicates. Hence these hypotheses could never increase the simplicity of any system to which we might add them. Indeed, they would inevitably add to the complexity of any system to which they were added, unless they were accompanied by such hypotheses as 'There are no entelechies,' 'There is no ether,' 'No World Soul exists,' or the like. And of course these latter hypotheses, although they nullify the mischief which the former hypotheses would do, at the same time make the former hypotheses pointless and uninteresting and show that we might just as well have done without these nonempirical terms altogether.

III

A final conclusion which we might draw if the proposed view of confirmation were to be accepted relates to the metaphysical status of the objects spoken of in empirical hypotheses. Many writers on the philosophy of science have chosen to suppose that there are no logical criteria in terms of which the truth or probability of scientific theories can be judged. They have supposed, in consequence, that scientific theories cannot be regarded as asserting anything about reality. Scientists frame theories in a spirit of more or less arbitrary caprice, the theories are no more than "postulates" or "stipulations," and the entities mentioned in them are "fictions" or "mental constructions." Theories may be "convenient" or "adequate," at best. Expressing this popular view, Schrödinger writes, "We prefer to say *adequate*, not *true*. For in order that a description be *capable* of being true, it must be capable of being compared *directly* with actual facts." [2]

If the viewpoint embodied in our proposed criterion of confirmation be accepted, then these Idealistic formulations must be rejected as misleading. If empirical knowledge really were confined to what can be directly observed, its span would be pitifully narrow indeed; and if all the rest of science were the tissue of fictions which these writers allege it to be, there would then be no intelligible reason for regarding the opinions of scientists as in any way superior to those of gypsies or spiritualists. These writers are wrong to draw an iron curtain between what is directly observable and what is not; for, if the argument we have been pursuing is at all correct, there are definite criteria in terms of which the degrees of confirmation of various empirical hypotheses and theories about unobserved things may be assessed. To assign to hypotheses various degrees of confirmation is to rate the degrees of their rational credibility; and to say that a hypothesis is rationally credible is to say that the evidence provides good reason for believing it to be true. Indeed, the principal thing to be said in favor of scientific hypotheses is that they are confirmed, that there is reason to believe them to be true. The assertion that the hypotheses which scientists propound are more fruitful or useful than those which can be obtained from old gypsy women would seem in the long run to be rather more doubtful.

---

[2] Erwin Schrödinger, *Science and Humanism* (Cambridge, 1951), p. 22.

Index

Index

Generalizations: inductive, 27, 43, 66, 69, 72, 87-88, 175, 182-183; involving numerical quantities, 92, 185-186
Goodman, Nelson, 29n., 174-176, 179, 188-190
Grue, 188
Gypsies, 12, 17, 19, 67, 200

Hamilton, Sir William, 162
Hempel, C. G., 133, 138, 148n., 154-155, 157n.
Hume, David, 10, 11, 108
Hypotheses, 2-3; concerning unobserved objects, 9, 95-99, 102, 105, 106-107, 114-115, 132, 152, 184-185; relational, see Relations; transcendent, 96
Hypothesis, method of, 153

Identity, 187-188
Indifference, principle of, 56
Induction: by elimination, 50-51; by simple enumeration, 50, 75, 78; concatenated, 79-80; problem of, 10, 13; see also Argument, inductive
Inference, see Argument
Infinite number of individuals, 69, 74-75
Infinite series, 65-66, 69, 104
Instance confirmation, 88-90
Intensional relations, 122

Johnson, W. E., 51

Kemeny, John, 85-87, 105n., 176, 178, 180
Keynes, J. M., 11, 51, 53, 55-60, 62, 68, 98-99, 102n., 160
Kneale, William, 67n., 129n.

Laplacean rule of succession, 88
Leprechaun, 117
Lewis, C. I., 34n., 122n.
Limited variety, 53, 55-56, 58-59

Logic, inductive vs. deductive, 5-6
Logical constructs, 109-110
Logical measure function, 86, 176, 178-179
Logical Positivism, 18
Logical structure, 129
Long run, 13-14, 68

McLendon, H. J., 130n.
Marksism, 130
Meaning postulates, 180
Metaphysical suppositions, 30, 69, 75, 139-142, 155-156, 159-160, 199
Mill, J. S., 63, 88
Models, 85-86, 176, 179

Nagel, Ernest, 15, 70n.
Newton, Isaac, 162
Nicod, Jean, 55

Observational statements, 102, 114, 181
Occam, William of, 162
Other minds, 6, 123-124
Over-arching Self, 155

Phenomenalism, 106-107
Phenomenology, 112
Physical objects, 8, 33, 107
Physicalistic statements, 36, 39, 42
Plato, 153
Popper, Karl, 156-161
Postulates: of a system, 168, 171; of scientific inference, 63-64
Pragmatism, 15, 17
Predicates: independence of, 84-85, 87; observational, 60, 96, 149, 190; purely qualitative, 189; theoretical, 96-97, 131, 139, 147, 196-197; two-placed, see Relations
Predictions, 46
Private language, 38
Probability, 32, 65, 70, 83-84, 86
Proper names, see Singular terms

Index